Alcohol Fuels

Also of Interest

*Renewable Natural Resources: A Management Handbook for the Eighties, edited by Dennis L. Little, Robert E. Dils, and John Gray

Energy from Biological Processes, Office of Technology Assessment, U.S. Congress

Energy Transitions: Long-Term Perspectives, edited by Lewis J. Perelman, August W. Giebelhaus, and Michael D. Yokell

*Energy Futures, Human Values, and Lifestyles: A New Look at the Energy Crisis, Richard C. Carlson, Willis W. Harman, Peter Schwartz, and Associates

*The Economics of Environmental and Natural Resources Policy, edited by J. A. Butlin

Energy, Economics, and the Environment: Conflicting Views of an Essential Interrelationship, edited by Herman E. Daly and Alvaro F. Umaña

Energy Analysis and Agriculture: An Application to U.S. Corn Production, Vaclav Smil, Paul Nachman, and Thomas V. Long II

Solar Energy in the U.S. Economy, Christopher J. Pleatsikas, Edward A. Hudson, and Richard J. Goettle IV

The Forever Fuel: The Story of Hydrogen, Peter Hoffmann

Agriculture as a Producer and Consumer of Energy, edited by William Lockeretz

*Available in hardcover and paperback.

Westview Special Studies in Natural Resources and Energy Management

Alcohol Fuels: Policies, Production, and Potential
Doann Houghton-Alico

Since the 1973 OPEC oil crisis, the rise of imported crude oil prices, and the questionable availability of petroleum supplies, the United States has been forced to investigate liquid-fuel alternatives. Alcohol fuels, including methanol and ethanol, offer the most realistic near-term potential as gasoline extenders or substitutes.

This book is the most current and complete source of information to date on alcohol fuels. Covering domestic policy and legislation, production processes, uses, and research and development, as well as international production and use, and presenting specific data--on, for example, the use of alcohol fuel in vehicles with spark-ignition and diesel engines, as a chemical feedstock, and in utility boilers and fuel cells--it is directed at research scientists, policymakers, and potential investors and producers. The appendixes include technical reference data, a bibliography, a list of information sources, and a glossary.

Doann Houghton-Alico, president of Technical Information Associates, Inc., is coauthor of Fuel From Farms: A Guide to Small-Scale Ethanol Production published by the Solar Energy Research Institute/ Department of Energy. Her previous positions include project consultant to government and research institutions, and public interest lobbyist and legislative analyst in Washington, D.C.

Alcohol Fuels
Policies, Production, and Potential
Doann Houghton-Alico

Westview Press / Boulder, Colorado

Westview Special Studies in Natural Resources and Energy Management

Copyright   1982 by Westview Press, Inc.

Published in 1982 in the United States of America by
    Westview Press, Inc.
    5500 Central Avenue
    Boulder, Colorado 80301
    Frederick A. Praeger, President and Publisher

Library of Congress Cataloging in Publication Data
Houghton-Alico, Doann.
    Alcohol fuels.
    (Westview special studies in natural resources and energy management)
    Bibliography: p.
    Includes index.
    1. Alcohol as fuel. I. Title. II. Series.
TP358.H68   338.4'7662669   82-2029
ISBN 0-86531-245-1   AACR2

Composition for this book was provided by the author
Printed and bound in the United States of America

To Daya, Chad and Peter

Hope is the thing with feathers
That perches in the soul,
And sings the tune without the words
And never stops at all....

                  Emily Dickinson

Contents

xi

Tables

Preface

Information presented in this book about the
alcohols--specifically, methanol, ethanol, and
butanol--ranges from history to state-of-the-art
technology, production to use, domestic policies to
international activities. The purpose is to examine the
potential of nonpetroleum-based liquid fuels and
chemical feedstocks. Alcohol fuels present the most
practical domestically available alternative to
petroleum-based products.
For several years there has been discussion about
depleting resources. But it wasn't until the 1973
Organization of Petroleum Exporting Countries (OPEC)
embargo and ensuing rising prices of crude oil that the
realities of such shortages hit the world, particularly
the American public. (The United States has an annual
per capita energy consumption rate of $332 \ 10^6$ Btu
compared to a world average of $60 \ 10^6$ Btu.) The
shortages in 1973 were politically and economically
motivated; nevertheless, they reflect what actual
resource shortages would be like.
Crude oil is one type of fossil fuel, which still
has an important role to play in world energy
development. However, fossil fuels are basically
nonrenewable. Enhanced recovery techniques and the
discovery of new fields does not change that fact. They
merely permit more time for the finding of
alternatives. The effect of political-economic power
plays must not be forgotten. One piece of the jigsaw
puzzle of world peace is most certainly predicated on a
continuing international supply of energy resources.
Alcohol fuels present a reasonable alternative.

Doann Houghton-Alico
Denver, Colorado

Acknowledgments

The author accepts full responsibility for the content of this book, but nevertheless wishes to recognize the important contributions made by others.

Jacqueline Sharkey, who served as editor, and Roxanne Oberst, who did the word processing, deserve particular thanks for their long hours and patient work.

Very special thanks go to Sherry Cole, Andrew Moriarity, Joe Nebolon, Dennis Orr, and Steve Rubin, who contributed in many ways.

I owe a particular debt of gratitude and respect to my technical reviewers for their areas of expertise: John Alico, Alico Engineers and Appraisers, Inc.; Harvey Blanch, University of California at Berkeley; Ron Bullard, Western Energy; Alicia Compere and William Griffith, Oak Ridge National Laboratory; David Locke, Queens College; Randall Noon, Kansas Energy Office; Robert Merten, Colorado Gasohol Promotion Committee; and Thomas Reed, Solar Energy Research Institute.

I would also like to acknowledge my research assistants, Marcy Dunning, Sandra Goldthorpe, and David Smythe.

Doann Houghton-Alico

1
Policy Issues

Overview

The stages of civilization are defined by their tools, e.g., the Stone Age, the Bronze Age. Tools are created by using available energy sources--such as human labor or fire--and natural resources. These tools are then used to further exploit the energy sources and natural resources to move the civilization to the next stage. Advances have occurred as a geometric progression, with the past century presenting an explosion of tools and technologies: today, machines exist that can perform a diverse and large amount of work; there are also machines to tell those machines what to do.

The costs and implications of this evolutionary process are difficult to evaluate. But the goal--across centuries and continents--has been economic growth and development. In the 20th century, the Western world has dominated development. The Western model, based on capitalist economics but comprising mixed economic theories, has brought unsurpassed wealth to the West. It is useful to remember, however, this definition of development: "Development is to be or to become. Not only to have." (Dadzie 1980.)

For much of this century, the achievements of the Western world--the First World--were the goals and expectations of the Third World--the underdeveloped countries. Two primary patterns have clouded that view in the past decade: (1) a questioning of the underlying values of the Western model, and (2) an inability of the world community to equitably distribute an adequate supply of essential resources. For example, in 1975, the average world rate of energy consumption was 2 k/person/yr of quasi-continuous power; the United States' rate was 11 kw/person/yr; the Third World rate was less than 1 kw/person/yr (Sassin 1980).

1

Those in yet-to-be-developed countries generally
continue to aspire to Western productivity. Many in the
West, however, have come to question the cost of that
productivity and its impact on the quality of life. It
is clear that unbridled development has brought more
than affluence. The other side of the coin is quite
tarnished--polluted air, water, and land. Now the
challenges are to bring together the goals of
development with the values of quality and continuity of
life and to ensure that such a lifestyle is available to
all people in all nations.

Two essentials for a nation's survival are energy
and food. Access to these commodities sets the stage
for economic growth and the basis for a stable society.
Shortages--real and imagined--of these resources
jeopardize international trade, development, and,
ultimately, world peace.

National self-reliance--not self-sufficiency--is a
concept that can be used to resolve food and energy
shortages and can be implemented in the developed as
well as the underdeveloped world. Self-reliance is a
concept based on domestic production of essential needs;
it does not preclude international trade. It requires
regional planning, cooperation among various segments of
society, and a commitment to national goals found in few
countries today.

Energy Resources

The energy crisis is not a matter of running out of
energy; that would defy the currently known laws of
physics. Rather, it is a shortage of a preferred form
of energy--petroleum derived from crude oil--resulting
from problems with supply, distribution, and control of
this resource.

Historically, energy development has involved the
discovery and exploitation of the cheapest, most
abundant, and most useful energy sources. The major
categories of energy use are lighting, heating, cooking,
transportation, and work performed by stationary
machines. Electricity and liquid fossil fuels are the
primary sources of energy for these uses.

For some time, fossil fuels--primarily coal and
crude oil--have been filling the world's diverse energy
needs. Fossil fuels are those fuels created when animal
and plant remains from prior geologic ages are
compressed in the earth's crust. They are basically
nonrenewable.

Fossil fuels frequently are the power behind the
switch; that is, they provide the fuel for electrical
power plants. Crude oil also provides the fuel for

transportation, the essential element in the mobile society of the developed world. (See Tables 1.1 and 1.2 at the end of this chapter for statistics on production and consumption of energy resources and statistics on consumption of energy by the transportation sector in the United States for 1980.)

Opinions differ widely on when the world might run out of fossil fuels, particularly crude oil. But the important point is that these fuels will run out, despite sporadic gluts resulting from temporarily stepped-up production and/or increased storage facilities. The only thing these respites accomplish is to buy more time to find alternative energy sources and enable nations to extend present reserves for strategic uses.

World crude oil reserves have been estimated at between 1,800 - 2,100 billion barrels. The United States' share is estimated at approximately 243 billion barrels. Enhanced recovery techniques may increase the United States' amount by 60 - 80 billion barrels (U.S. Department of Energy 1978). (See Table 1.3 at the end of this chapter for statistics on production, imports, and cost of crude oil in the United States.) The cost in additional energy for enhanced recovery will have to be carefully evaluated. An additional problem with domestic supplies is that they comprise heavy, high-sulphur, low-quality crude oil, and U.S. refineries are designed to handle the lighter imported crude. Standard Oil Co. of California is spending $1 billion to improve a 280,000 bbl/day refinery in Pascagoula, MS. The improved refinery will be able to handle heavier crude oil, but capacity will not be increased. Approximately 80% of the world's reserves are high-sulphur crude oil, which is difficult to refine into gasoline, especially unleaded gasoline. The National Petroleum Council estimates that $5 billion - $12 billion will have to be spent on refinery improvements by 1990 in order to use the crude oil that will be available (Mufson 1981).

The lack of extensive petroleum reserves is exacerbated by the increased use of petroleum. In 1960, world energy consumption was approximately 150 quadrillion Btu/year, or 150 quads, of which petroleum provided 51 quads. In 1990, world energy consumption is expected to reach 400 quads, with petroleum supplying 160 quads, an increase of just over 300% in 30 years. In 1970, total United States energy consumption was 67,143 trillion Btu; petroleum consumption was 29,537 trillion Btu, or 44.0%. In 1975, United States consumption was 70,707 trillion Btu; petroleum consumption was 32,731 trillion Btu or 46.3%. In 1978, United States consumption was 78,154 trillion Btu;

petroleum consumption was 37,965 trillion Btu, or 48.6%
(U.S. Department of Energy 1980).

Global energy supplies and needs are a major
consideration in international relations. Crude oil is
supplied to the world market by relatively few
countries; therefore, economic and political
manipulations can become a major international problem.
Of the 1,675,423 thousand metric tons of crude oil
exported in 1979, 1,342,328 thousand metric tons were
exported by the Organization of Petroleum Exporting
Countries (OPEC) (Statistical Office, Department of
International Economic and Social Affairs, United
Nations 1981). In 1980, OPEC comprised Algeria,
Ecuador, Gabon, Indonesia, Iran, Iraq, Kuwait, Libya,
Nigeria, Qatar, Saudi Arabia, United Arab Emirates, and
Venezuela. (See Table 1.4 at the end of this chapter
for international exports, imports, and per-capita
consumption of crude oil for 1979 listed by country.)

Potential political upheaval and the resulting
vulnerability of supplies is possible. For example,
13 of the present Arab heads of state came to power
forcibly; 12 wars have been fought within the Arab
nations since 1965; and 60% of the exports from the Arab
states bordering the Persian Gulf pass through three
ports, with the oil controlled at eight critical pump
sites (Levy 1980).

In 1973, OPEC announced an embargo on crude oil
exports followed by a major price increase; a "shot"
that clearly has been felt around the world. The
effects have been to emphasize to the general public the
finiteness of this particular resource, to bring an end
to development based on cheap energy, and to alter
dramatically the international economic order.

In addition to political, economic, and resource
availability problems with fossil fuels, environmental
problems have also been created. A report in 1977 by
the National Academy of Sciences warned of the
"greenhouse effect" caused by excessive release of
carbon dioxide (CO_2) from the burning of fossil fuels.
Major climate changes and shifts in agricultural zones
affecting crop production could result (U.S. Council on
Environmental Quality and U.S. Department of State
1980).

Issues

The primary sources of liquid fuels are
nonrenewable fossil fuels. Relying on these fuels for
the future is short-sighted at best. This does not
imply that they don't have a role to play, but as these
fuels are being used, time should be spent to develop
alternatives.

Alcohol fuels provide a proven alternative. Comparisons with other domestically available liquid fuel resources are limited. However, in addition to the alcohols, certain vegetable oils can be used as liquid fuels. The research is encouraging, but more information is needed before the role they can play in meeting future energy needs is fully evaluated.

The issues that need to be examined to evaluate alcohol fuels as alternatives to petroleum-based liquid fuels include:

1. availability of feedstocks;
2. viability of production processes;
3. environmental impact of use of feedstocks and of production processes;
4. performance in specified uses, such as motor vehicle fuel in spark ignition engines;
5. national priorities for feedstock use and product use; and
6. energy balance.

The first three issues are discussed in the chapters on the individual alcohol fuels--methanol, ethanol, and butanol. The fourth is discussed in the chapter on alcohol fuel use.

In the case of ethanol and butanol, which are produced from agricultural crops, setting priorities for feedstock use is an important issue.

The problem of world hunger is complex. It is not a matter of running out of food or the ability to produce food, although production, storage, and distribution patterns contribute to the problem. Essentially it is an issue of control over food production resources, including land, seeds, tools, water, and credit. For example, export agriculture, advocated by many governments, international organizations, and economists in theory, may have an adverse impact on world hunger. For a developing country, raising agricultural exports may take up arable land needed for domestic foodstuffs, and may make it more difficult for the rural and urban poor to acquire food. For many nations, perpetual famine is the result.

The nature of famine has changed in the past few decades. Throughout history, famine was associated with disastrous events, such as major floods, droughts, or prolonged wars. Today, however, famine is a continual experience for a large part of the world's population. Providing sufficient calories--estimated by the U.N. Food and Agriculture Organization as 2,354 calories per capita per day--is only part of the problem. It is also essential to provide adequate protein. Analysis of world food needs has indicated that one major problem is food distribution--particularly of protein-- internationally, within nations, and within households.

Carefully formulated policies geared to national self-reliance are necessary to combat the problems of world hunger. Technology alone does not solve problems; rather, it is a tool that can be used to contribute to solutions. The Green Revolution, an attempt at a technological solution to the world's food problem through the creation of hybrid strains, didn't work. It solved some problems, and created others, for example, the water and pesticides needed for growing the hybrids taxed water resources and created additional environmental problems.

Agricultural crops can be used to produce fuel without a depletion of food resources. It is reasonable to establish a basic criterion that fuel production from agricultural crops should not reduce the amount of food currently marketed or needed in the future for increasing populations. Given this criterion, there are several approaches to alcohol production, for both the near- and long-term. These approaches include:

1. converting feed crops with minimal loss of feed value;
2. using fuel-crop production on idle land ill-suited to food-crop production;
3. using excess production;
4. using spoiled or perishable produce;
5. using distressed or marginal crops;
6. using cellulose agricultural wastes;
7. cultivating high-yield fuel crops.

A restructuring of American agricultural practices to produce both food and fuel crops is possible, and should include consideration of crop diversity; cultivation of high-yield fuel crops; land husbandry, including crop rotation and soil conservation; and emphasis on regional and integrated (production of both livestock and crops) farming operations.

Integrated farms have the flexibility to grow the varieties of crops necessary to competitively produce alcohol over the long term and can use the stillage (the high-protein residue from the production process) directly as animal feed. This efficient land use may help revive the economic viability of such farms. It is quite possible that small-scale ethanol plants with a centrally located plant for ethanol drying and blending with gasoline may make the most sense economically, especially if the stillage can be used locally, without drying, as an animal feed. Regionally located, larger-scale butanol plants may have the same advantages.

In the developing world the issue of food versus fuel takes on different dimensions. Some believe that the coordination necessary to produce both is impossible and, therefore, production of alcohol fuels should be

avoided because it will inevitably lead to a decline in
essential food production. While it is true that such a
scenario is possible, it is not the only conceivable
outcome. Small-scale production of ethanol or larger-
scale production of regionally located methanol
or butanol plants have the potential of improving the
standard of living for the rural population in many
developing countries. The key is controlling and
allocating resources. If large-scale plantations,
particularly under foreign ownership, are implemented as
fuel farms, small-scale and subsistence farmers will be
dislocated and join the urban migration with its tragic
consequences--joblessness, increased crime, breakdown of
community and family values. Locally owned cooperatives
can provide an alternative and a structure for planning
and development to ensure the production of crops needed
for food and fuel, with the proper use of recoverable
coproducts. Using domestic resources to satisfy
domestic needs is a major step towards self-reliance.
Although potential export crops may be decreased, so
will the increasingly expensive and ultimately
unavailable crude oil imports.

Another crucial issue is the energy balance of
energy production. Energy balance is essentially adding
up the amount of one form of energy that is necessary to
produce another form of energy and then comparing the
totals. It is important to understand that any effort
to convert one fuel form to another results in a net
loss of available energy. Electrical energy provides an
excellent example. Electricity is one of the most
convenient forms of energy: it can be almost instantly
transported great distances; large amounts of energy can
be transmitted using systems that occupy relatively
little space; the uses of electricity are varied.
But the production of electricity exacts a price in
available energy. If it is produced in a coal-fired
power plant, 10 Btus of coal must be burned to yield
3.33 Btus of electrical energy, which means a net energy
loss of 6.66 Btus, or 66.6%. In addition, another 3% is
lost in transmission and distribution (U.S. Department
of Energy 1981). This decidedly negative energy
balance is acceptable because of the tremendous
convenience and superior quality of electrical power.
For example, a coal fire does produce light, but the
quality of light is markedly inferior to a single
100-watt light bulb. Also, coal by itself cannot
produce much of the work that electricity can produce.
Therefore, consumers pay the price.

Liquid fuels are another example. They are needed
for strategic as well as day-to-day uses. Therefore,
consumers are going to have to pay a price in lost
energy to produce usable liquid fuels from other

resources. In the early days of automobile design, some vehicles had their own wood burners. That obviously is not practical today. But wood can be used as a feedstock to produce methanol, which, in turn, can be used as a fuel in automobiles.

In determining energy balances, many factors are considered. In comparing energy balances, it is essential to know what those factors were, otherwise equal entities may not be compared. Specific energy balances for methanol, ethanol, and butanol are discussed in their respective chapters.

One means of evaluating the potential role of alcohol fuels in a comprehensive energy policy is to compare the amount of liquid fuels used for alcohol fuel production with the amount of liquid fuels produced. The liquid fuel energy balance is rarely discussed in liquid fuel production statistics. With ethanol, for example, the liquid fuel requirements for production and transportation of dry land corn are 6.96 gal/acre (Wittmuss 1975). The average yield per acre for corn is 75 bushels, from which 187.5 gallons of ethanol can be produced. Liquid fuels are generally not used in the fermentation-distillation process. This means 0.037 gallons of liquid fuel are consumed in crop production and transportation for every gallon of ethanol produced. To state this figure another way, only one gallon of liquid fuel is required to produce 27 gallons of ethanol. This represents a net liquid fuel gain of 26 gallons.

LIQUID FUELS

Fuels are substances that produce useful power when burned. There are five types of fuels: (1) solid; (2) liquid; (3) gas; (4) atomic; and (5) chemical, sometimes referred to as exotic fuels. Solid, liquid, and gas are the common types of fuels and comprise primarily carbon and hydrogen, which, when reacted with oxygen in the air under specific conditions, give off heat and create new substances.

Liquid fuels have several advantages. They are (1) easy to handle, (2) convenient to store, (3) easy to transport, and (4) contain few impurities. Crude oil derivatives and the alcohols are the most common liquid fuels.

In 1970, the United States imported 483.3 million barrels of crude oil at a cost of $2.76 billion, or $5.71/bbl. In 1978, the United States imported nearly 2.3 billion barrels at a cost of $39.1 billion, or $17/bbl. In 1980, the United States paid $33.89/bbl for imported crude oil. Posted prices in January 1981 were $37.59/bbl (Oil and Gas Journal 1980; U.S. Department of Commerce 1979, 1980; U.S. Department of Energy 1981).

Crude oil can be used to produce various products. The following list gives the products and respective number of gallons derived from one barrel of crude oil, which is 42 gallons:

1. gasoline, 18.32 gal;
2. distillate fuel oil, 9.40 gal;
3. residual fuel oil, 5.05;
4. jet fuel, 2.77;
5. petrochemical feedstocks, 1.51;
6. still gas, 1.51;
7. processing loss, 1.51;
8. asphalt, 1.21;
9. coke, 1.05;
10. liquefied gases, .97;
11. kerosene, .51;
12. lubricants, .50;
13. miscellaneous, .42;
14. special napthas, .26;
15. road oil, .04;
16. wax, .04; and
17. ethane, .04 (Rounds 1981).

ALCOHOL FUELS

Alcohols are a large and diverse family of chemical compounds. Alcohols are hydroxyl derivatives of hydrocarbons, identified by their unique arrangement of carbon, hydrogen, and oxygen atoms. The three classifications of alcohols are primary, secondary, and tertiary.

Primary alcohols have one hydroxyl group--the carbon-hydrogen combination attached to the carbon in the carbinol group (C-OH). Secondary alcohols have two hydroxyl groups attached to the carbinol carbon, and tertiary alcohols have three hydroxyl groups attached. (See Table 1.5 at the end of this chapter for diagrams of sample primary, secondary, and tertiary alcohols.)

Polyhydroxyl alcohols are those with more than one hydroxyl group. An example of this is ethylene glycol (1,2-ethanediol). (See Table 1.6 at the end of this chapter for a diagram of the structure of ethylene glycol.)

Alcohols are primarily used as chemical feedstocks--raw materials for the production of other products. Nevertheless, they are liquid fuels and can be used as substitutes for petroleum-based liquid fuels in various applications. The higher alcohols--those with the most carbon atoms--have the highest heating

value because more carbon-hydrogen bonds can be broken
to produce more energy. The higher alcohols are
generally more difficult and expensive to produce,
however. Therefore, attention is focused primarily on
the simpler alcohols--methanol and ethanol--with one and
two carbon atoms, respectively. Butanol, with four
carbon atoms, also shows promise as an alternative to
petroleum-based liquid fuels and can be produced by a
fermentation process that appears to be reasonable in
terms of cost and technology required.

Examples of alcohols and their uses include:

1. Methanol (CH_3OH)--also called methyl alcohol
or wood alcohol--is used as a motor vehicle
fuel and a chemical feedstock in industry.

2. Ethanol (C_2H_5OH)--also called ethyl alcohol or
grain alcohol--is the basis for the alcoholic
beverage industry, and is used as motor
vehicle fuel, most commonly in a blend with
gasoline known as Gasohol. Gasohol is a
registered trade name for 10% agriculturally
derived fermentation ethanol and 90% unleaded
gasoline.

3. N-butanol (C_4H_9OH)--1-butanol or butyl
alcohol, commonly referred to as butanol--is
used as a solvent for lacquers, resins, and
adhesives; in hydraulic fluids, antibiotic
recovery, and urea-formaldehyde resin; as
amines for gasoline additives; and to produce
butyl acetate (McCutchan 1954). Butanol also
has potential as a motor vehicle fuel.

4. Isopropyl alcohol (C_3H_7OH)--also called
2-propanol or rubbing alcohol--is used for
medical purposes and to produce acetone.

5. Ethylene glycol ($H_4C_2_2OH$)--also called 1,
2-ethanediol--is used as an antifreeze. This
is a secondary alcohol with two hydroxyl
groups.

6. Glycerol ($C_3H_5_3OH$)--also called glycerin--is
used as an antifreeze, in alcoholic beverages,
in pharmaceutical products, and to make
nitroglycerine. This is a tertiary alcohol
with three hydroxyl groups.

For the purpose of differentiating alcohols used
for fuels and feedstocks from beverage alcohol, the term
alcohol fuels is used in this book. The emphasis is on
examining alcohol fuels as alternative liquid fuels,
although their importance as chemical feedstocks is
discussed.

Table 1.1
Production and Consumption of Energy Resources in the
United States for 1980[1]

Type of Energy	Production by Quads	Consumption by Quads
Coal	18.877	15.603
Crude Oil	18.250	34.196 (Petroleum)
Natural Gas, Plant Liquids	2.263	–
Natural Gas, Dry	19.754	20.495
Hydroelectric Power	2.913	3.125
Nuclear Electric Power	2.704	2.704
Geothermal and Electricity from Biomass	0.114	0.114
Total Energy	64.876	76.201[2]

[1]Numbers are independently rounded.
[2]Includes a difference of 0.037 for coal coke exports above production and imports.

Source: U.S. Department of Energy 1981.

Table 1.2
Consumption of Energy by the Transportation Sector in
the United States for 1980

Type of Energy	Consumption by Quads
Natural Gas, Dry	0.606
Petroleum	17.987
Electricity, Sales	0.011
Total	18.604[1]

[1]Does not include electrical energy losses attributed to this sector.

Source: U.S. Department of Energy 1981.

12

Table 1.3
United States Crude Oil Production, Imports, and Cost

Year	Domestic Production (thousand bbl/day)	Imports (thousand bbl/day)	Cost[1] US$/bbl Domestic	Cost[1] US$/bbl Imported
1976	8,132	5,287	8.84	13.48
1977	8,245	6,594	9.55	14.53
1978	8,707	6,195	10.61	14.57
1979	8,552	6,452	14.27	21.67
1980	8,597	5,177	24.23	33.89

[1]Refiner acquisition cost.

Source: U.S. Department of Energy 1981.

Table 1.4
Imports, Exports, Per-Capita Consumption[1]
of Crude Oil for 1979[2,3]

Country	Imports (Thousand Metric Tons)	Exports (Thousand Metric Tons)	Per-Capita Consumption (kg)
AFRICA			
Algeria	350	47,500	167
Angola	--	6,500	145
Congo	--	2,000	401
Egypt	--	4,350	290
Ethiopia	600	--	20
Gabon	--	7,000	3,676
Ghana	1,200	--	108
Ivory Coast	1,600	--	202
Kenya	2,350	--	166
Liberia	500	--	277
			continued

[1]Per-capita consumption reflects in-country use, including refinery use and is not descriptive of the actual crude oil equivalent used per-capita.
[2]The 1979 numbers are based on preliminary figures. A final update is to be available in late 1981.
[3]Only countries with crude oil activity are listed. Those countries with no domestic supplies and no refineries are not included in this table.

Table 1.4, continued

Country	Imports (Thousand Metric Tons)	Exports (Thousand Metric Tons)	Per-Capita Consumption (kg)
AFRICA, continued			
Libyan Arab Jamahiriya	N/A	92,800	2,276
Madagascar	500	--	59
Morocco	3,400	--	176
Mozambique	425	--	42
Nigeria	--	111,000	44
Senegal	750	--	136
Sierra Leone	355	--	105
Somalia	350	--	91
South Africa	13,000	--	436
Sudan	850	--	48
Togo	960	--	388
Tunisia	1,239	5,235	249
United Rep. of Tanzania	500	--	28
United Rep. of Cameroon	--	1,747	--
Zaire	180	920	10
Zambia	900	--	159
ASIA/SOUTH PACIFIC			
Australia	8,500	200	2,074
Bangladesh	1,200	--	14
Brunei	--	11,880	563
Burma	--	40	39
China	--	9,500	102
Democratic Kampuchea	-0-	--	--
Guam	1,546	--	13,328
India	14,700	--	42
Indonesia	1,270	61,570	158
Japan	239,672	--	2,287
Korea Dem. People's Rep.	1,600	--	91
Korea, Republic of	25,013	--	669
Malaysia	2,500	9,000	473
Mongolia	6	--	4

continued

Table 1.4, continued

Country	Imports (Thousand Metric Tons)	Exports (Thousand Metric Tons)	Per-Capita Consumption (kg)
ASIA/SOUTH PACIFIC, continued			
New Zealand	2,500	--	966
Pakistan	3,500	--	50
Philippines	9,400	--	226
Singapore	34,221	324	14,345
Sri Lanka	1,350	--	92
Thailand	8,456	11	183
EUROPE			
Albania	-0-	--	786
Austria	8,612	--	1,395
Belgium	33,990	250	3,477
Bulgaria	13,000	--	1,481
Czechoslovakia	19,000	250	1,237
Denmark	8,676	100	1,717
Finland	10,510	--	2,311
France incl. Monaco	124,000	--	2,356
German Democratic Rep.	20,000	--	1,200
Germany, Fed. Rep. of	109,280	--	1,809
Greece	17,780	1,400	1,608
Hungary	10,500	1,200	1,059
Ireland	2,380	--	702
Italy incl. San Marino	114,861	--	2,019
Netherlands	60,446	--	4,303
Norway	6,320	15,860	2,199
Poland	16,220	--	472
Portugal	8,482	--	842
Romania	14,000	--	1,193
Spain	48,300	--	1,303
Sweden	16,221	145	1,957
Switzerland incl. Liechtenstein	4,633	--	730
United Kingdom	60,384	42,264	1,724
Yugoslavia	11,500	--	706

continued

Table 1.4, continued

Country	Imports (Thousand Metric Tons)	Exports (Thousand Metric Tons)	Per-Capita Consumption (kg)
MIDDLE EAST			
Bahrain	9,932	--	42,842
Cyprus	530	--	849
Iran	--	119,215	922
Iraq	--	160,070	509
Israel	8,300	--	2,173
Jordan	1,473	--	463
Kuwait incl. Pt. Neut. Zone	--	106,973	14,639
Lebanon	1,855	--	601
Oman	--	14,680	-0-
Qatar	--	23,848	2,200
Saudi Arabia	700	443,000	3,654
Syrian Arab Republic	2,240	2,060	1,095
Turkey	11,600	--	317
United Arab Emirates	--	89,132	797
Dem. Yemen	1,950	--	1,061
NORTH AMERICA			
Canada	30,673	14,025	3,806
Mexico	--	35,065	612
United States of America	319,659	11,522	3,276
SOUTH AND CENTRAL AMERICA			
Antigua	--	--	--
Argentina	1,728	--	974
Bahamas	9,700	--	43,304
Barbados	140	--	701
Bolivia	210	150	240
Brazil	50,041	1,500	463
Chile	3,750	--	400

continued

Table 1.4, continued

Country	Imports (Thousand Metric Tons)	Exports (Thousand Metric Tons)	Per-Capita Consumption (kg)
SOUTH AND CENTRAL AMERICA, continued			
Colombia	1,350	--	298
Costa Rica	430	--	194
Cuba	6,350	--	647
Dominican Republic	1,400	--	249
Ecuador	-0-	6,220	545
El Salvador	760	--	163
Guatemala	800	--	119
Honduras	448	--	119
Jamaica	1,050	--	481
Martinique	450	--	1,429
Netherlands Antilles	30,150	1,000	112,115
Nicaragua	610	--	243
Panama	2,400	--	1,287
Paraguay	225	--	76
Peru	-0-	3,650	384
Puerto Rico	12,000	--	3,438
Trinidad and Tobago	7,388	6,267	10,40
U.S. Virgin Islands	30,675	--	280,952
Uruguay	2,250	--	782
Venezuela	--	74,000	3,676

Source: Statistical Office, Department of International Economic and Social Affairs, United Nations 1981.

Table 1.5
Sample Compositions of Primary, Secondary,
and Tertiary Alcohols with Common Name,
Classification, and Scientific Name

$$CH_3-\overset{\overset{\displaystyle H}{|}}{\underset{\underset{\displaystyle H}{|}}{C}}-OH \qquad CH_3-CH_2-\overset{\overset{\displaystyle OH}{|}}{CH}-CH_3 \qquad CH_3-\overset{\overset{\displaystyle CH_3}{|}}{\underset{\underset{\displaystyle CH_3}{|}}{C}}-OH$$

ethanol	2-butanol	2-methyl-2-propanol
primary alcohol	secondary alcohol	tertiary alcohol
ethyl alcohol	sec-butyl alcohol	tert-butyl alcohol

Table 1.6
Composition of Ethylene Glycol

$$\begin{array}{l} H_2C-OH \\ | \\ H_2C-OH \end{array}$$

REFERENCES

Chase Manhattan Bank. 1981. The Petroleum Situation. Vol. 5, no. 1-2. New York, NY: Chase Manhattan Bank, N.A.

Dadzie, K.K.S. 1980. "Economic Development." Scientific American. Vol. 243, no. 3; September. New York, NY: Scientific American, Inc.

Lappe, F.M., Collins, J. 1978. Food First. New York, NY: Ballantine Books.

Levy, Walter J. 1980. "Oil and the Decline of the West." Foreign Affairs. Summer. New York, NY: Council on Foreign Relations.

Mufson, Steve. 1981. "Oil Concerns Have Too Many Refineries, Yet Too Few To Handle Heavier Crude." Wall Street Journal. 18 March. New York, NY: Dow Jones and Company, Inc.

Oil and Gas Journal. 1980. "OPEC posted prices." April.

Rounds, Michael. 1980. "It's A Long Journey from Oil Wellhead to Gasoline Pumps." 10 February. The Rocky Mountain News. Denver, CO.

Sassin, Wolfgang. 1980. "Energy." Scientific American. Vol. 243, no. 3, September. New York, NY: Scientific American, Inc.

Solar Energy Research Institute. 1980. Fuel From Farms: A Guide to Small-Scale Ethanol Production. Golden, CO: Solar Energy Research Institute.

Statistical Office, Department of International Economic and Social Affaris, United Nations. 1981. 1979 Yearbook of World Energy Statistics. New York, NY: United Nations.

U.S. Council on Environmental Quality and U.S. Department of State. 1980. The Global 2000 Report to the

President: Entering the Twenty-First Century. Vol 1.
Washington, D.C.: Government Printing Office.

U.S. Department of Commerce. 1979. U.S. Foreign Trade:
USA Commodity by Country, Annual. Washington, D.C.:
U.S. Department of Commerce.

U.S. Department of Commerce. 1980. Current Business
Statistics. May. Washington, D.C.: U.S. Department of
Commerce.

U.S. Department of Energy. 1978. "Enhanced Recovery of
Oil and Gas." Washington, D.C.: U.S. Department of
Energy, Office of Public Affairs. Document OPA-009
(8-78).

U.S. Department of Energy. 1980. Monthly Energy
Review - October. Washington, D.C.: U.S. Department of
Energy, Energy Information Administration. Document
DOE/EIA-0035 (80/10).

U.S. Department of Energy. 1981. Monthly Energy Review
- June. Washington, D.C.: U.S. Department of Energy,
Energy Information Administration. Document
DOE/EIA-0035(81/06).

Winston, S.J. n.d. Ethanol Fuel: Use, Production
Principles and Economics. Golden, CO: Solar Energy
Research Institute.

2
Methanol

PRODUCTION

Properties of Methanol

Methanol--also referred to as methyl alcohol or
wood alcohol--is the simplest compound in the class of
organic compounds called alcohols. The formula is
CH_3OH; methanol is two molecules of hydrogen chemically
liquefied by one molecule of carbon monoxide. The
hydroxyl group--OH--in methanol imparts polarity to the
molecule. Methanol is colorless, odorless, and miscible
in water in all proportions (Reed 1974). It has a
molecular weight of 32.04, freezes at -97.8°C (-208°F),
boils at 64.5°C (148°F), and has a flash point of 12.2°C
(54°F), a specific gravity of 0.79, an autoignition
temperature of 464°C (867.2°F), a Research Octane Number
of 106, and a density of 6.59 lb/gal at 20°C (68°F)
(Hawley 1981).

History of Production and Use

Historically, methanol was used as a lighting fuel,
replacing whale oil about 1830, only to be replaced by
kerosene about 1880. In the late 1800s, methanol was
used for heating and cooking. Starting about 1920,
methanol was used as a solvent, as a feedstock in
plastics manufacturing, and for fuel injection in
aircraft piston engines in the United States. Because
of gasoline shortages in World War II, methanol was used
as a motor vehicle fuel (Reed 1974).
Methanol was originally produced from the
destructive distillation of wood, a process that also
yielded the coproducts of charcoal, acetic acid, and
acetone. With this process, the weight of the methanol
produced amounted to about 4% of the weight of the wood

used. In 1923, the Germans built a synthetic methanol plant that reacted carbon monoxide (CO) and hydrogen (H) at high temperature and pressure. In 1924, they were exporting methanol to the United States at two-thirds the cost of wood-distilled methanol. In 1926, the United States began using this process (Paul 1978).

Current Production Processes

Most U.S. methanol production starts with a synthesis gas (syngas) formed by gasification of fossil fuels with their sulphur compounds removed to ensure the catalytic reaction. Gasification is a thermochemical conversion of an organic feedstock into liquid fuel. See Table 2.1 for a description of the syngas reactions.

Table 2.1
Syngas Reactions from Natural Gas and Coal

CH_4 Natural Gas	+	H_2O Steam $\longrightarrow$	$CO + 3H_2$. Syngas
CH Coal	+ 0.5 O_2	$\xrightarrow{\text{Steam}}$	$CO + 0.5\ H_2$. Syngas
Source: Reed 1981a.			

Natural gas is the most common feedstock, although coal is also used. The gas is passed over a catalyst such as zinc oxide with either chromium or copper oxides, at pressures ranging from 50 - 350 atm and temperatures as high as 400°C (752°F). The water-gas shift reaction starts with steam and carbon, derived from a source such as coal. This produces CO and H_2. This mixture, called syngas, is put through the Fischer-Tropsch process for synthesizing hydrocarbons to produce carbon dioxide (CO^2) and hydrocarbons. The ratio of the gases is changed to approach a stoichiometric ratio of 2:1. This technology is best suited for large plants. The production rate is a function of the equilibrium concentration of methanol in the reacting gases and the rate at which this equilibrium can be approached. This, in turn, is a function of the composition of the gas; the catalyst used; and the temperatures, pressures, and rates at which the gases contact the catalyst (Paul 1978).

In the original high-pressure gasification process, pressures of 300 atm at 200°C (392°F) were used in the presence of a zinc-chromium oxide catalyst. In 1968 a low-pressure process was developed using 50 atm at 250°C (482°F) with a highly-selective, copper-based catalyst. This process produces purer methanol. Since that time, several processes have been developed using intermediate pressures (Reed 1973).

Major U.S. methanol producers and plant capacities are given in Table 2.2.

Table 2.2
Major U.S. Methanol Producers and Plant Capacities

Producers	Plant Capacities 10^6kg
Du Pont-Beaumont, TX	67
Orange, TX	33
Celanese-Bishop, TX	17
Clear Lake, TX	77
Borden-Geismar, LA	53
Georgia Pacific- Plaquemine, LA	33
Monsanto-Texas City, TX	33
Hercules-Plaquemine, LA	33
Tenneco-Pasadena, TX	27
Rohm & Haas-Deer Park, TX	7
Air Products-Pace, FL	17
CSC-Sterlington, LA	17
Total	414

Source: Paul 1978. Reprinted by permission of Noyes Data Corporation.

When natural gas is the feedstock, excess hydrogen is produced, which then can be used to produce ammonia. Natural gas is often burned off in oil refining facilities because of problems with storage and distribution. Methanol-ammonia plants at these sites could utilize this currently wasted natural gas, rather than depleting natural gas resources for home heating and other essential uses.

Potential Production from Biomass

There is another way to produce methanol besides the inefficient destructive distillation of wood and the synthetic method using nonrenewable fossil fuels as the original feedstock. Biomass--plant and animal matter, wastes, and residues that have the potential of being converted into usable energy resources--provides an alternative feedstock for gasification production of

methanol. Generally, biomass has a low sulphur
content--0.1% to 0.2%--compared to coal-->2%.
    Two primary ways exist for producing methanol from
biomass. The first is anaerobic digestion of wet
biomass (e.g., sewage, manure, algae) to produce
methane, which in turn can be used to produce methanol,
or can be used as a gas itself, although it is not a
practical motor vehicle fuel. The second is partial
combustion of dry biomass (e.g., wood, agricultural and
municipal wastes) to produce syngas (CO + H_2O), which in
turn can be used to produce methanol. This latter
process is described in Table 2.3.

Table 2.3
Methanol Production from Biomass

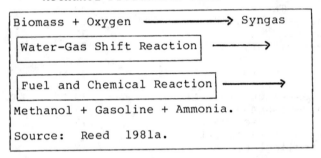

Biomass + Oxygen ————————> Syngas

Water-Gas Shift Reaction ————————>

Fuel and Chemical Reaction ————————>

Methanol + Gasoline + Ammonia.

Source: Reed 1981a.

    The water-gas shift reaction corrects the H:CO
ratio for methanol synthesis. It is described in Table
2.4.

Table 2.4
Water-Gas Shift Reaction

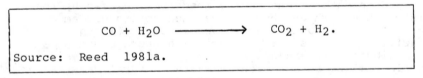

$$CO + H_2O \longrightarrow CO_2 + H_2.$$

Source: Reed 1981a.

    The production of methanol from biomass involves
the following steps: (1) biomass preparation, (2)
gasification, (3) gas compression, (4) CO shift, (5) CO_2
removal, and (6) methanol synthesis.
    By mid-1981, no commercialized biomass syngas
gasification process using oxygen was in operation.
However, the Solar Energy Research Institute (SERI) has
developed a prototype gasifier. Air cannot be used,
because nitrogen, which is present in air, prevents
methanol from being produced. High-grade methanol is
necessary for use as a chemical feedstock and solvent,

but lower- grade methanol (which also contains ethanol, propane, and isobutanol in a ratio of approximately 2:3:5) may be used as a motor vehicle fuel (Reed 1981b). Vulcan Cincinnati, Inc. uses the trade name "Methyl-fuel" for this lower-grade methanol.

Table 2.5 describes the syngas reaction from biomass, and Table 2.6 describes the fuel and chemical reactions from syngas.

Table 2.5
Syngas Reaction from Biomass

$$CH_{1.4}O_{0.6} + 0.2O_2 \longrightarrow CO + 0.7H_2.$$
Biomass   Oxygen   Syngas

Source: Reed   1981a.

Table 2.6
Fuel and Chemical Reactions from Syngas

$CO + 2H_2 \longrightarrow CH_3OH.$
Methanol

$CO + 3H_2 \longrightarrow CH_4 + H_2O.$
Methane + Water

$CO + 2H_2 \longrightarrow CH_2.$
Gasoline

$N_2 + 3H_2 \longrightarrow NH_3.$
Ammonia

may occur at same plant

Source: Reed   1981a.

FEASIBILITY CONSIDERATIONS

To evaluate the feasibility of using domestically produced methanol as an alternative to petroleum liquid-fuel products, it is necessary to determine the following:
1.   availability of feedstocks;
2.   impact of feedstock use for methanol production on other feedstock uses;
3.   capital and operating costs;
4.   environmental impact of feedstock choice and production process; and
5.   energy balance (energy used for energy gained).

Feedstocks

Potential feedstocks for methanol production are natural gas, coal, and biomass. Rising prices and priority uses such as home heating make natural gas a poor choice of feedstock for methanol production. However, natural gas that is burned off as waste in the petroleum refining process could be captured and used for methanol production. Coal appears to be a poor long-term feedstock choice because it is essential for power generation and its mining and use pose environmental problems. It could provide a near-term feedstock, however. Both coal and natural gas are nonrenewable fossil fuels, and further exploitation will reduce their availability and lessen the time for developing alternatives.

Biomass--particularly in the form of wood and agricultural and municipal wastes--appears to be the best long-term feedstock for methanol production. Silvaculture--the science of tree farming--has developed rapidly in the last several years, particularly as a result of increased knowledge of plant genetics, and has made wood a more available feedstock. Agricultural and municipal wastes are readily available in many locations. A principal advantage of using waste material as a feedstock is that it saves money and results in less environmental degradation than using other feedstocks.

Investment and Operating Costs

Capital and operating costs of methanol production depend on the feedstock, site, and production process. Because of the complex, three-stage process used for methanol production--syngas production, adjustment through the water-gas shift reaction, and reaction to produce methanol--large-scale, capital-intensive plants are needed. A 430 ton/day (43 Mgal/year) plant advertised for sale in June 1981 was appraised at $22 million, which is 51¢/gal capacity. The plant is located in Texas. Included in the cost were truck- and rail-loading facilities, maintenance facilities, and 1,146 acres of land. Lignite coal fields are located in the area.

The present cost of methanol production is misleading when considering methanol as an alternative motor vehicle fuel. The reason is that methanol, at least through 1981, has been used primarily in industry where high-grade methanol has been required. However, high-grade methanol is not required for motor vehicle fuel use, and the cost of producing lower-grade methanol is expected to be considerably less.

Production Factors

A potentially practical and cost-effective process for producing methanol as an alternative motor vehicle

fuel appears to be using biomass as a feedstock in an
oxygen gasifier to produce syngas. The syngas is then
reacted to produce methanol and ammonia. In mid-1981
such a plant did not exist, therefore actual costs
cannot be analyzed.

The cost of producing methanol from biomass is
influenced by:

1. feedstock availability and cost;
2. conversion-plant size;
3. the biomass production and conversion process
    chosen;
4. potential reduction in capital-cost
    requirements as a result of technological and
    process improvements;
5. economic incentives; and
6. process-energy optimization (Wan 1979).

The following developments could greatly improve
the economic competitiveness of methanol produced from
biomass:

1. large-scale, low-cost biomass production from
    energy farms;
2. coproduct conversion systems; e.g., methanol
    and ethanol, methanol and acetic acid,
    methanol and ammonia, and methanol and
    gasoline (Wan 1979);
3. packaged methanol-ammonia plants utilizing
    agricultural, municipal, and forest wastes;
4. new catalysts to permit methanol production at
    lower pressure and temperature;
5. methanol reactors in liquid media;
6. direct production of methanol from methane;
    and
7. improvements in other forms of production,
    such as pyrolysis (Reed 1981a).

SERI has developed a pressurized biomass oxygen
gasifier. The goal is to produce a clean $CO-H_2$ syngas
at 10 atm to use in commercial gas pipelines, for
turbine generation, and to produce methanol and
ammonia. Specific goals for the SERI gasifier are:

1. to achieve high temperature (900° - 1000°C)
    (1652° - 1832°F) equilibrium in the bed to
    crack tars and hydrocarbons;
2. to maintain the bed temperature with staged
    oxygen injection;
3. to achieve rapid bed pyrolysis with an oxygen
    flame; and
4. to use 10 atm pressure for increased
    throughput, improved equilibrium, and cost
    reduction for downstream purification and
    pumping (Reed 1981b).

Eight test runs were completed between October 1980
and January 1981. The gas analysis of test runs 6 and 8
is shown in Table 2.7. The feed rate was approximately

20 lb/hr, pellet size was 1/8 in, and steam-oxygen was
used in run 8 to produce a higher H_2:CO ratio gas.
Oxygen consumption was 0.4 ton/ton of biomass; oxygen
costs ranged from $0.50 to $1.50/MBtu of biomass
processed; and the methanol-ammonia value was $10/MBtu.
Plans call for building a full-scale prototype in
1983-1985, using 200 tons of biomass/day and producing
120 tons of methanol and ammonia/day. However, changes
in government funding priorities may affect these plans.

Table 2.7
SERI Gas Analysis for Oxygen Gasifier - Runs 6 and 8

Gas	Run 6	Run 8
H_2	18.5	28.9
CO	55.4	20.3
CO_2	10.9	44.0
N_2	14.3	3.6
CH_4	0.9	1.5
$>>O_2$	TR	0.4
H_2:CO Ratio	0.33	1.38

Source: Reed 1981b.

Market Potential

Methanol may be used as a fuel or a chemical
feedstock for industry. As a fuel, methanol may be used
in internal combustion (IC) engines, turbine engines,
utility boilers, and in fuel cells. It is possible that
methanol used in fuel cells may need to be of a higher
purity than methanol used in engines and boilers.

Methanol's high octane ratings, which are
indicative of its high-performance combustion
capabilities, make it an outstanding fuel for IC
engines. Its performance is increased in stratified
charge engines. Tests indicate straight methanol
provides leaner operation and fewer pollutants than
gasoline. A higher fuel injection rate should be used
with methanol than with gasoline, and problems with cold
starting and corrosion must be resolved. Other
considerations include (1) adequate lubrication and
cooling, (2) inert or protected fuel-distribution
components, and (3) spark plugs with adequate heat
dissipation to prevent preignition.

Methanol may be used in blends with gasoline up to
15% - 20% with no engine alterations. However,
efficiency can be improved with engine modifications.
Problems include corrosiveness, phase separation with

water, vapor lock, and cold starting. Solutions to these problems are being or have been developed.

Diesel engines require a fuel with a cetane rating of 45 - 55, indicating a low autoignition temperature. Methanol has a cetane rating of 3 and a correspondingly high octane rating. Dual-fuel operation is possible using a diesel fuel for ignition and methanol for the remainder of the charge. Tests on this application at the University of Minnesota showed lower carbon particulate emissions, lower intake manifold temperature, unchanged efficiency, and unchanged unburned emissions (Paul 1978).

Methanol may provide an alternative or additive to petroleum-based motor vehicle fuels. The decision is essentially more political than technical. There are some problems and some advantages. The problems can be resolved, but government priorities will influence whether or not methanol is used on a large-scale basis in the transportation sector in the United States.

Methanol appears to be an excellent fuel for turbine engines. Technical problems present no major difficulties. Tests have been run using methanol as a turbine fuel by General Electric Co. and United Technologies, Inc. While methanol may be used in unmodified turbines, the following modifications are recommended to maximize use:

1. additional lubrication;
2. increased fuel flow rates; and
3. explosion proofing.

In addition, the sulphur content of the fuel must be kept below 1 ppm, because of problems with corrosion. Methanol typically contains <0.1 ppm sulphur. Methanol's miscibility with water, which may contain sodium salts, could present a problem unless there are adequate safeguards during storage and distribution. Such salts can cause corrosion during the high temperature operation necessary for most turbines (Paul 1978). Obstacles to its use are supply and cost. Results of recent tests comparing methanol with gasoline included:

1. increased thermal efficiency;
2. reduced nitrogen oxide (NO_x) emissions; and
3. increased CO emissions (Paul 1978).

Methanol may also be used as a fuel for electrical utility boilers. Tests were made by Vulcan Cincinnati, Inc. in 1972 on multifuel boilers modified with Y-shaped burner nozzles for complete combustion and a centrifugal pump to replace fuel-oil pumps in order to handle the low viscosity and low lubricity of methanol. The results were:

1. no particulates were emitted;
2. soot deposits from previously burned petroleum fuels were burned off;

3. there was less radiative and more convective heat transfer, causing 3% lower boiler efficiency than with natural gas at equal loads;
4. no sulphur compounds were emitted;
5. CO emissions were lower than with natural gas;
6. aldehyde emissions were less than 1 ppm; and
7. acid emissions were less than 10 ppm (Paul 1978).

The conclusions were that methanol could serve as an excellent boiler fuel and that the major obstacles to its use were cost and availability.

Research indicates that methanol may be used directly or indirectly in fuel cells. One advantage is that methanol can be catalytically converted to syngas at low temperatures [about 250°C (482°F)]. Also, methanol forms no carbon, contains no sulphur (because it must be removed before synthesis), and is in convenient liquid form. Tests using methanol in fuel cells have been or are being conducted in the United States by: the U.S. Army, National Aeronautics and Space Administration, DOE, General Electric Co., General Motors Corp., Exxon Corp., Union Carbide Corp., United Technologies, Inc., and Monsanto Corp. Tests abroad have been or are being conducted by: AIG, Telefunken, Siemans A.G., K.V. Kordesch, and Battelle Institut e.v. in Germany; Electrical Research Association and Shell Oil Co. in England; Institut Francais du Petrole and the Alsthom Co. in France; and Hitachi, Ltd. in Japan (Paul 1978).

Methanol also has been used commercially as a feedstock for growth of single-cell protein. Although methanol is toxic if consumed by humans, most animals, and many plants, it has proven an effective medium for growth of certain single-cell algae, bacteria, and yeasts. In Europe, such products grown on a methanol medium are replacing milk and soybeans in calf feed, and have innumerable potential uses in the world protein market (Paul 1978).

Environmental and Safety Considerations

Combustion of methanol as a motor vehicle fuel has advantages over combustion of gasoline. Methanol can replace lead as an octane booster. Replacing lead as a gasoline additive is mandated by the U.S. Environmental Protection Agency. Hydrocarbon emissions are lower than with gasoline, and NO_x and CO emissions can be controlled. Sulphur oxides and sulphuric acid are not present, nor are carbonaceous particulates. Aldehyde emissions are increased, however, and unburned fuel (methanol) emissions are present (Paul 1978).

Large-scale methanol spills pose less of an
environmental hazard than petroleum spills, but adequate
safety precautions must be taken. Methanol's solubility
in water is an advantage and disadvantage. It prevents
methanol from being cleaned up in the same manner as
gasoline in an aquatic spill; but water could be used to
clean up methanol on non-porous surfaces. Pseudomonas
fluorescens, a bacteria, consumes methanol and may be
used for cleanup of aquatic spills. Methanol is toxic
to most aquatic plants and organisms, although no more
toxic than gasoline (Paul 1978; Moriarity 1977).

Explosion and combustion are not as serious a
hazard as with gasoline. Safety precautions include
(1) use of inerts to reduce flammability limits, (2)
separation of oxidizing agents, (3) enclosure to prevent
evaporation, and (4) storage away from heat or the sun.
Water spray, carbon dioxide, and dry chemical or foam
extinguishers will put out methanol fires. Fire
extinguishing with water spray is an advantage of
methanol over gasoline.

Methanol is sometimes fatal if ingested by humans.
Because of its name--methyl alcohol--and its association
with grain alcohol, many cases of methanol poisioning
have been documented. Toxicity symptoms include
(1) headache, (2) mucous membrane irritation,
(3) nervousness, (4) trembling, (5) nausea, (6)
blindness, and (7) vomiting. See Table 2.8 for toxicity
of methanol, benzene, and gasoline.

Exposure to methanol fumes in enclosed locations is
safe to 200 ppm, the established Threshold Limit Value.
Reports in the past several years have discussed
possible dangers of methanol toxicity externally. A
review of the literature by Dr. Andrew Moriarity of
Biomedical Resources International in Toronto, Canada,
indicates that the incidents cited in these reports are
from 1904, 1910, 1913, 1915, and 1954, and resulted from
wood-alcohol poisoning. Wood alcohol--methanol from the
destructive distillation of wood--contains significant
amounts of acetone and other ingredients. It is
unreasonable to attribute these cited cases directly to
methanol. Methanol is readily available in products
such as Sterno, charcoal lighter fluid, windshield
washer antifreeze, model airplane glue, and denatured
alcohol. Additional cases of external methanol
poisoning have not been reported, therefore, it appears
that this is not a problem today. Nevertheless, with
increased methanol use, potential toxicity should be
monitored (Moriarity 1976).

Problems with materials wear include corrosion.
Corrosion from methanol use has been noted with lead
plating (terneplating), lead, zinc, aluminum, magnesium,
copper, and some plastics, such as Viton. Steel is not
affected (Paul 1978).

Table 2.8
Comparative Toxicity Ratings[1]

	Eye Contact	Inhala-tion	Skin Pene-tration	Skin Irri-tation	Inges-tion
Methanol	2	2	2	1	1
Benzene	2	4	2	2	2
Gasoline	(2)	(3)	(3)	(1)	(2)

[1] 1 = mild; 5 = extreme toxicity; ( ) = estimated-- depends on composition.

Source: Paul 1978. Reprinted by permission of Noyes Data Corporation.

Energy Analysis

As with the production of ethanol, mediating factors should be considered when analyzing the energy balance of methanol.

Two important factors are the need for and availability of liquid fuels. More than 40% of U.S. imported crude oil is used for vehicular transportation. It is clear, therefore, that alternative liquid fuels or significant liquid-fuel extenders can provide part of the solution to immediate and long-term domestic energy problems.

Even though coal converts to methanol efficiently, there are overriding reasons why coal should not be used as a long-term feedstock for methanol production. These include its other priority uses, the environmental problems resulting from its mining and use, and the fact that it is a nonrenewable resource.

To evaluate accurately the energy balance for methanol as a liquid fuel, the energy cost of producing the feedstocks and the energy required for methanol production must be known. Such numbers are difficult to extrapolate from available data on methanol research and production.

The potential energy resource from a given feedstock, assuming a 60% efficiency, can be coded as shown in Table 2.9.

Table 2.9
Potential Energy From Feedstocks

$$\frac{\text{Mass}}{\text{Ratio}} = \frac{\text{Btu/Ton Methanol}}{.6(\text{Btu/Ton Feedstock})} = \frac{\text{Tons Feedstock}}{\text{Ton Methanol}}$$

Table 2.10 uses this formula to show the amount of feedstock it would take to produce one ton of methanol at 60% energy efficiency.

Table 2.10
Feedstock Conversion to Methanol

Feedstock	10^6 Btu/Ton	Tons Feedstock to Produce 1 Ton Methanol (60% Efficiency)
Coal	15-30	1.1-2.2
Lignite	12	2.8
Wood, Agricultural wastes	16	2.1
Municipal wastes	9	3.7
Source: Reed 1981a.		

Research indicates that methanol is an efficient fuel, capable of meeting federal controlled-emission standards with little reduction in conversion efficiency in vehicle use. Final evaluations on the role of methanol in a multifuel energy policy require additional information on production costs and efficiencies (Hagen 1977).

34

REFERENCES

Hagen, David. 1977. "Methanol as a Fuel: A Review with
Bibliography." Warrendale, PA: Society of Automotive
Engineers. SAE/PT-80/19.

Hawley, Gessner G., ed. 1981. The Condensed Chemical
Dictionary, 10th ed. New York, NY: Van Nostrand
Reinhold Co.

Moriarity, Andrew J. 1976. "Testimony before the
California Senate Committee on Public Utilities, Transit
and Energy." Toronto, Canada: Biomedical Resources
International.

Moriarity, Andrew J. 1977. "Toxicological Aspects of
Alcohol Fuel Utilization." Toronto, Canada: Biomedical
Resources International.

Moriarity, Andrew J. 1981. Biomedical Resources
International, Toronto, Canada. Personal communication.

Paul, J.K. 1978. Methanol Technology and Application
in Motor Fuels. Park Ridge, NJ: Noyes Development Corp.

Posner, Herbert S. 1975. "Biohazards of Methanol in
Proposed New Uses." Journal of Toxicology and
Environmental Health. No. 1, pp. 153-171.

Reed, Thomas B., Lerner, R.M. 1973. "Methanol: A
Versatile Fuel for Immediate Use." Science. Vol. 182,
28 December. Washington, D.C.: American Association for
the Advancment of Science.

Reed, Thomas B., Lerner, R.M. 1974. "Sources and
Methods for Methanol Production." Cambridge, MA:
Massachusetts Institute of Technology, Energy
Laboratory, Methanol Center.

Reed, Thomas B. 1975. "Biomass Energy Refineries for
Production of Fuel and Fertilizer." Cambridge, MA:
Massachusetts Institute of Technology, Energy
Laboratory, Methanol Center.

Reed, Thomas B. 1981a. Solar Energy Research
Institute, Golden, CO. Personal communication.

Reed, Thomas B.   1981b.   "SERI High Pressure Oxygen
Biomass Gasifier Status."   Golden, CO: Solar Energy
Research Institute.

U.S. Department of Energy.   1978.   Fuels from Biomass
Program.   Washington, D.C.: DOE.   Stock No.
061-000-00063-1.

Wan, E.I., et al.   1979.   "Biomass-Based Methanol
Processes."   Proceedings from the Third Annual
Biomass Energy Systems Conference Proceedings, 5-7 June
1979.   Golden, CO: Solar Energy Research Institute.

3
Ethanol

PRODUCTION

Properties of Ethanol

Ethanol--also referred to as ethyl alcohol, grain alcohol, and EtOH--is a primary alcohol with the formula CH_3-CH_2-OH (or C_2H_5OH).

Ethanol is colorless, volatile, and miscible in water in all proportions. It has a molecular weight of 46.1, boils at 78.3°C (172.9°F), freezes at -117.3°C (-243.1°F), and has a density of 0.789 at 20°C (68°F), a calorific value of 12,800 Btu/lb, a latent heat of vaporization of 204 Cal/g, and a Research Octane Number reported in the range of 91 to 105 (Paul 1979; Hawley 1981; Timmermans 1965; Winston n.d.b).

History of Production and Use

Ethanol was probably "discovered" accidentally thousands of years ago, when the intoxicating effect of the spontaneous fermentation of sugar was noticed. Fermented beverages are mentioned in the Rig Veda, an ancient sacred book of India; the Old Testament of the Bible; and early Egyptian and Arabic records. A comprehensive text, Das Kleine Distillierbuch, was published in Strasbourg, France, in 1500.

In 1808, the first continuous still was built in France. Because ethanol was also used as a beverage, it was heavily taxed. These taxes presented an economic burden, which made ethanol unfeasible for industrial use until the denaturing process was adopted. The first denaturant law was passed in Great Britain in 1855. It provided for the addition of a substance (the denaturant) to ethanol to make it unfit for human consumption. This exempted ethanol from excise taxes (Paturau 1969). In the United States in the mid-1800s, ethanol replaced whale oil in lamps because it was clean and odorless. The use of ethanol as an industrial chemical increased in the late 1800s. In the 1920s and 1930s, straight alcohol (also called neat or pure alcohol) fuels or alcohol (usually ethanol)-gasoline blends were used in Argentina, Australia, Cuba, Japan, New Zealand, the Philippines, South Africa, and Sweden. During this period, Chrysler Motor Corp. produced cars modified for use of 100% alcohol fuel for export to New Zealand, and International Harvester Co. produced trucks to run on straight ethanol (also called E100) for export to the Philippines. World Wars I and II brought increased use of ethanol. During World War II, a blend of ethanol and gasoline was marketed in the Midwest under the name "Agrol." Ethanol was used in synthetic rubber production, as an airplane and submarine fuel, and in torpedoes. In 1945, the United States was producing nearly 1.1 billion proof gallons of ethanol in 200 plants. After the war, interest in ethanol as a fuel declined, and little was done about using ethanol blended with gasoline until 1971. At that time, the combination of rising fuel prices and the need for new markets for agricultural products prompted the state of Nebraska to pass legislation reducing their gasoline tax for Gasohol (a blend of 90% unleaded gasoline and 10% agriculturally derived anhydrous ethanol). The oil embargo in 1973, and the continually rising price and uncertain supply of crude oil have resulted in states, institutions, and government agencies exploring various aspects of developing alcohol fuels, including ethanol (Stark 1954; Paul 1979).

Fermentation Process Overview

Ethanol may be produced by (1) fermentation of sugar, and (2) chemical reaction produced by acid-catalyzed hydration of ethylene. Ethylene is produced as a by-product of petroleum refining and by the dehydration of ethanol. Ethanol produced from ethylene is not described here, since one purpose of this book is to examine alcohol fuels as alternatives to petroleum-based products.

The fermentation process for producing ethanol is established technology and is the basis of the grain-alcohol beverage industry. The established production process gives ethanol a head start in the race for domestically available nonpetroleum liquid fuels. The existence of an established grain-alcohol fermentation technology does not mean, however, that this process is the most efficient for today's needs. Plants designed to produce grain-alcohol beverages focus on a product for human consumption, whereas plants designed for alcohol fuel production can be more concerned with energy efficiency. Plant design

modifications and production process modifications are being examined. Developments under study include:
1. high ethanol-tolerant yeast strains to reduce separation costs;
2. thermophilic yeast strains to reduce fermentation times;
3. biological production of other alcohols and chemicals;
4. new uses for the stillage coproduct;
5. continuous fermentation-distillation;
6. vacuum fermentation;
7. improved methods for breaking the azeotrope, including codistillation with gasoline; and
8. other separation methods, including freezing, absorption, and membranes to lower separation energy use and costs (Reed 1980).

The basis of fermentation ethanol production is the specific chemical change undergone by the substrate, which is induced by an enzyme or microorganism. This reaction is summarized in Table 3.1 with glucose as the substrate.

Table 3.1
Fermentation of Glucose

$$C_6H_{12}O_6 \xrightarrow{\text{enzyme}} 2C_2H_5OH + 2CO_2 + \text{Heat}$$

Glucose   Ethanol   Carbon dioxide

Fermentation Production Stages

The stages of fermentation ethanol production are:
1. feedstock storage;
2. substrate preparation;
3. fermentation;
4. distillation;
5. ethanol drying;
6. ethanol storage; and
7. coproduct treatment.

Feedstocks for ethanol production through fermentation are carbohydrates, which are compounds of carbon, hydrogen, and oxygen with a ratio of hydrogen to oxygen of 2:1. Carbohydrates are most commonly formed by photosynthesis. Sugars (saccharides) are low molecular weight carbohydrates composed of one or more saccharose group(s). The simplest are monosaccharides. A hexose is a monosaccharide composed of six carbon atoms. Glucose and fructose are two hexoses capable of being fermented. Although they have the same formula-- $C_6H_{12}O_6$ --their molecular structure is different as demonstrated in Table 3.2 (Flowers n.d.; Hawley 1981).

40

Table 3.2
Molecular Structure of Glucose and Fructose

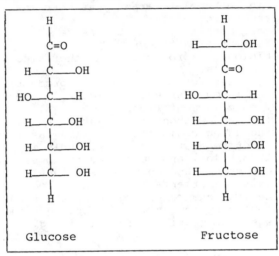

Glucose   Fructose

In the more complex sugars--the disaccharides and polysaccharides--the monomeric units are joined together by a chemical bond called the glycolytic link. The process of hydrolysis is used to break this link and change the complex sugars into fermentable monosaccharides.

Feedstock Storage. For efficient fermentation, the feedstocks must be properly stored and prepared. Storage systems must be used that can protect against carbohydrate or sugar loss.

Once the feedstock(s) is (are) chosen, carbohydrate or sugar losses can be minimized by identifying potential loss mechanisms and designing an appropriate storage system to eliminate these. Loss mechanisms include spoilage, sprouting, and freezing.

Spoilage and sprouting of cereal grains can be minimized by keeping them relatively dry. Storage time for grains varies with moisture content and grain temperature. Therefore, partial drying may be necessary at time of harvest to prevent immediate losses.

Tuber and root crops with high moisture content spoil easily and should be processed quickly or dehydrated if they are to be stored. Potatoes must be stored in cool, dark facilities to prevent germination. Freezing potatoes does not affect carbohydrate content. Pesticides containing arsenic compounds have been used for many years on potato crops. Molds such as Aspergillus glaucus, Aspergillus virens, and Mucor

<u>mucedo</u> grow readily on potato starch, and in the
presence of even trace quantities of arsenic they
produce a toxic compound, the gas trimethylarsine.
Fermentation conditions are ideal for propagation of
these <u>Aspergillus</u> species, and production of hazardous
gases could result (Winston 1980). There is no
documentation of this situation presenting serious
problems.

Juice from sugar crops is the most difficult
feedstock to store because of its high moisture and
sugar content, which is conducive to contamination. A
variety of bacteria and fungi use sugar as a primary
food source. The rate of bacterial action is affected
by the osmotic pressure. In a solution with a high
sugar content, there is a spontaneous tendency for water
to pass through the cell membrane into the solution.
The driving force for this is manifested as the osmotic
pressure. The result is the bacterium becomes
dehydrated and dies (Locke 1981). Methods for
concentration of the juice are (1) evaporation, (2) heat
sterilization, (3) chemical sterilization, and (4)
ultrafiltration. There are attendant equipment and
processing costs (Paul 1979).

<u>Substrate Preparation</u>. Conversion of the feedstock
into a substrate capable of being fermented includes
cleaning, preparation for sugar extraction, and sugar
extraction to provide a 20% - 24% concentration of
simple sugars in an aqueous solution.

Grains may be cleaned by blasts of air or screening
to remove stones and debris; root crops may require
high-pressure streams of water; sugar beets and
Jerusalem artichoke tubers need to be washed.

Grains are milled in roller, hammer, or attrition
mills. The type of mill utilized depends on the
feedstock used and particle size desired. Hammer mills
are generally preferable (Bullard 1981). The particle
size is determined by the cooking procedure to be used.
The existence of flour in the grind may decrease alcohol
yield, because the flour particles may become scorched
before the larger particles are thoroughly gelatinized.
Therefore, milling to the point of producing flour must
be avoided. Other starch crops, such as potatoes and
cassava, may be sliced or ground.

After milling, grinding, or slicing, the
long-branched starch chains are converted into glucose.
This may be done by acid or enzymatic hydrolysis.
Hydrolysis is a chemical reaction involving water and
another substance, in which the water molecule is
ionized, and the compound hydrolyzed is split (Stark
1954; Hawley 1981).

Acid hydrolysis is a one-step process, in which the
feedstock is mixed with a mild solution of sulphuric

acid. One problem with acid hydrolysis is that the process may continue and consume some of the glucose produced. In this case, maximum potential yields are not realized.

Enzymatic hydrolysis takes place in two steps: liquefaction--also called dextrinization or, more commonly, cooking--and saccharification. First, the prepared feedstock is mixed with water to form a slurry, and the alpha-amylase--the liquefying enzyme--is added. The mixture is agitated continuously, which assists in breaking down the starch. The slurry is heated at temperatures high enough to cause the cells to absorb water and expand. This action is called gelatinization. Optimal gelatinization is essential. Too little and complete liquefaction cannot take place; too much and the enzyme cannot get to the starch molecules. The liquefying enzyme breaks down the starch molecules into dextrins, a form of sugar. The efficiency of liquefaction is a function of temperature, residence time at the appropriate temperature, the rate of heat-up, the uniform distribution of heat, and the amount of available liquefying enzyme (Stark 1954; Winston 1980; Bullard 1981).

For saccharification to begin, the temperature and pH must be changed and a different enzyme added. Saccharification breaks down the dextrins into glucose, a fermentable sugar. The conditions of pressure, temperature, and pH that enable the enzymes to work most effectively during saccharification vary from one enzyme to another.

Dehydrated tuber and root crops may be treated in a manner similar to grains. The high moisture content of fresh crops must be taken into consideration if cooking them directly. Fungal amylase has been used for saccharification of potatoes, sweet potatoes, and cassava (Stark 1954). There are indications that putting potatoes through the cooking cycle twice will yield a more easily fermentable substrate. The pH should be approximately 6 (Bullard 1981).

Wet milling is a method incorporated into substrate preparation that separates out the protein, fiber, and starch in the original grain. The starch is then converted into sugar to make alcohol, and the protein and fiber may be treated as coproducts. The protein in this case is suitable for human consumption. Because of the high initial costs for wet milling, it is feasible only for large-scale plants (Reed 1980).

The process involves the following steps:

1. the grain is soaked in an aqueous solution of very dilute sulphur dioxide;
2. the hulls are ripped apart by metal grinding wheels without mashing the germ;

3. the germ is separated from the rest of the grain in a centrifuge;
4. the material is passed over a screen to separate the starch and protein from the fiber and hulls;
5. the starch and protein are blended with water to make a slurry; and
6. the slurry is sent to a centrifuge, which separates the protein (gluten) from the starch (Reed 1980).

Preparing sugar crops for storage by extracting the juice also accomplishes substrate preparation. Molasses, a by-product of sugarcane refining, may be used the same way as sugarcane juice, and requires no further substrate preparation (Stark 1954; Solar Energy Research Institute 1980).

<u>Fermentation.</u> Although fermentation is only one step in the production process, it is the key step. Fermentation is a biochemical process in which enzymes produced by microorganisms transform an organic substance (substrate) into ethanol. The substrate used in ethanol fermentation is a 6-carbon sugar. The microorganisms used in fermentation—whether yeasts, molds, or bacteria—contain no chlorophyll, and therefore cannot produce their own food through photosynthesis. Rather, they produce enzymes that act as catalysts to convert carbohydrates from organic material into energy. These enzymes start the fermentation process and the biochemical changes that produce ethanol from the sugar substrate. The microorganisms used are facultative anaerobic—they can grow in both anaerobic and aerobic conditions.

Yeasts are the most commonly used microorganisms responsible for producing the enzymes that convert glucose to ethanol. The complex reactions and interrelationships of yeast fermentation of a sugar substrate is called the Embden-Meyerhoff-Parnas pathway. <u>Saccharomyces</u> is the yeast genus generally used in ethanol production. Other yeast genuses and other non-yeast microorganisms may also be used (Stark 1954).

The primary factors that affect production yield and efficiency are:
1. physiological condition of the inoculum—the microorganisms added to the mash;
2. environmental factors present during fermentation; and
3. the quality of the substrate (Stark 1954).

The physiological condition of the inoculum is dependent on optimal growth conditions, which vary depending on the specific microorganism used. The most important environmental conditions are pH and temperature. Other factors are (1) buffer capacity, (2) initial load of contaminants, (3) mash concentration, (4) alcohol concentration, (5) selection of yeast strains, (6) nutritional requirements of the yeast, and (7) amount of oxygen present (Stark 1954). Some agitation may assist in yeast efficiency (Bullard 1981).

In mash that has been enzymatically hydrolyzed, a pH above 5.0 favors lactic acid formers; and a pH below 4.1 results in the inactivation of the amylases. The optimal pH for residual dextrin conversion is 4.8 - 5.0. After saccharification, pH adjustment may be accomplished by the addition of sulphuric or lactic acids. If the mash were hydrolyzed with acid, the addition of hydrated lime could be used to adjust the pH. If molasses is used as the feedstock, the pH is usually set at 4.8 - 5.0 (Stark 1954; Casida 1968).

Selection of the fermentor temperature is governed by (1) fermentor volume, (2) mash concentration, and (3) probable atmospheric conditions for the first 48 hours of fermentation (Stark 1954).

The fermentation process itself gives off heat of about 500 Btu/lb of ethanol produced (Solar Energy Research Institute 1980). This must be taken into consideration to maintain the correct temperature for fermentation. The goal is to maintain as high a temperature as possible without reducing ethanol yield, in order to maintain an efficient fermentation rate.

It is possible to recycle the yeast. Benefits of recycling the yeast are:

1. the 3% - 6% sugar loss during cell formation is reduced, which happens in the batch fermentors as the yeast propagates;
2. equipment and vessel requirements for preparation of yeast inoculum are reduced;
3. fermentation time is reduced;
4. the cost of additional yeast is reduced;
5. the growth of contaminants is inhibited because the ethanol is formed more rapidly; and
6. less fouling of the distillation column occurs.

The Melle-Boinot process, used in Brazil's national alcohol industry, treats the recovered yeast with an acid at 2.5 - 3.0 pH for one to three hours to clean the inoculum and kill bacterial contaminants. The Usines-de Melle process uses centrifugal separators to obtain the yeast before the fermented mixture is distilled (Agricultural Alcohol Blended Fuel Study Conference 1975).

Maximum fermentation activity generally occurs between 12 - 30 hours after initiation and is

dependent on temperature, agitation, and yeast strain.
With temperature control, the fermentation process takes
approximately 40 - 60 hours (Stark 1954).

Microbial contamination can be minimized by adding
a large inoculum to the mash. This ensures that yeast
growth exceeds that of contaminants, and that the yeast
consume the available nutrients. A total liquid
inoculum can be from 5% - 8% of the mash. Sterilized
mashes are not nearly as subject to competitive enzyme
reactions from microbial contaminants; therefore, the
yeast inoculum need not be as large. Unwanted microbes
can also be controlled by using commercially available
antiseptics. Less expensive chemicals, such as sodium
pentachlorophenolate, aspergillic acid, ammonium
thiosulfate, acrinol, sulphur dioxide, tartaric acid,
tyrothricin, polymixin, and chloramphenicol may also be
used to control the growth of microorganisms without
affecting the yeast. Although not wholly reliable,
especially in advanced contamination cases, fusel oil
from previous distillations can be added to the mash to
inhibit microbial growth. Esteraldehydes, solutions of
0.01% formaldehyde, or mixtures of hydrofluoric acid or
sodium fluoride with the mash can be used to control
growth of lactic acid-forming bacteria (Winston n.d.;
Solar Energy Research Institute 1980). Formaldehyde,
hydrofluoric acid, and sodium fluoride in solution are
highly toxic by inhalation to humans. Extreme care must
be taken if using these chemicals. If antiseptics and
disinfectants are used, it may not be possible to sell
the resulting stillage for animal feed (Paturau 1969).

The selection of the yeast or other microorganism
is important in ethanol fermentation. Yeast selection
should be based on (1) rapid growth, (2) high alcohol
and sugar tolerance, (3) efficient conversion rate, (4)
a maximum growth temperature of at least 32.2°C (90°F),
and (5) hardiness to environmental conditions, such as
pH, temperature, and osmotic pressure. The optimal
temperature for yeast growth is 28.9° - 32.2°C (84° -
90°F). Temperatures above 32.2°C (90°F) inhibit yeast
activity and favor the growth of bacteria contaminants
(Stark 1954).

The sugar concentration of the mash is governed by
two basic concerns:
1. excessively high sugar concentrations
   inhibit the propagation of yeast cells in
   the initial stages of fermentation; and
2. high ethanol concentrations are lethal to
   yeast.

If the concentration of ethanol in the solution
reaches levels high enough to kill the yeast before all
the sugar is consumed, the quantity of sugar that
remains is wasted. Yeast propagation problems can be

overcome by using large inoculations to start
fermentation. Saccharomyces strains can utilize
effectively all the sugar in solutions that are 16% -
24% sugar, while producing a beer (fermented mash) that
ranges from 8% - 12% ethanol by volume (Solar Energy
Research Institute 1980). Lower percentages do occur.
    In the batch fermentation process, a specific
volume of the mash is fermented at one time. The
process continues until maximum yields are obtained.
Fermentation is then stopped, the products recovered,
the equipment cleaned and sterilized, and another batch
started. Most commercial fermentations--as of
mid-1981--are successive batch fermentations using
several fermentors. In the continuous fermentation
process, additional mash is added continuously or, more
commonly, at short intervals to the fermentor. The
advantages and disadvantages of batch and continuous
fermentations are shown in Tables 3.3 and 3.4.

Table 3.3
Batch Fermentation Advantages and Disadvantages

Advantages
1.   control of microbial contaminants;
2.   control of ethanol quality per batch.

Disadvantages
1.   high capital requirement for large-scale
      production because of the number of fermentors
      required;
2.   decrease in volumetric efficiency of
      fermentors, based on time used;
3.   possible variations from batch to batch,
      requiring homogenization.

Source: Righelato 1970.

    Distillation. Because the ethanol concentration in
the fermented beer is 6.5% - 12% by volume, it is
necessary to concentrate the ethanol produced to use it
as fuel. This is usually accomplished by distillation
(Stark 1954; Schroder 1981).
    Distillation is a separation process of two or more
liquids in solution that is based on their relative
volatilities and takes advantage of their different
boiling temperatures. Fractional distillation draws off
vapors from different levels of the distillation
column. Heavier products will be recovered in the lower
part of the column, and lighter products in the upper

part. Although the desired end product may be recovered in concentrations greater than those of additional by-products, it is not necessarily pure. The separation factor is the means for measuring the efficiency of separation. The separation factor is the ratio of the purity of one product stream to the purity of the second product stream (Hawley 1981; Winston n.d.b). For example, if one product stream contains 1000 units of ethanol and the second product stream contains 10 units of ethanol, the separation factor is 100.

Table 3.4
Continuous Fermentation Advantages and Disadvantages

Advantages
1. high volumetric efficiency;
2. yeast recycling;
3. establishment of a flow growth-rate equilibrium;
4. a more consistent product than can be obtained from batch fermentation;
5. lower capital and labor costs;
6. reduced time requirements.

Disadvantages
1. high potential for serious microbial contamination;
2. in spite of providing a more consistent product, homogenization is still required.

Source: Righelato 1970.

Distillation is accomplished in a multistage system composed of two or more columns. Each column purifies the ethanol further. As the process continues, the ethanol concentration in the liquid decreases, so the boiling temperature of the mixture increases. As the boiling temperature increases, the relative amount of water that evaporates also increases. The process of boiling and condensation is repeated many times, and each step extended until nearly all the ethanol from the previous step has been exhausted (Schroder 1981).

Rectification, also called fractionation, achieves further separation by allowing some of the vapor that has been condensed to flow back down the column. This downward flowing liquid is called the reflux. As the reflux comes into contact with the rising vapor improved product separation occurs.

The columns may use packing if the solids have been removed from the beer. If not, plate columns are used. In most two-column systems, the first is a plate column, and the second is a packed column (Bullard 1981).

Packed columns utilize an inert material to baffle the downward flow of the reflux; plate columns utilize a series of plates for the same purpose. Proceeding up the column, each plate would have lower pressure, lower temperature, and more alcohol (Schroder 1981; Hawley 1981).

Approximately 95% ethanol may be achieved through distillation. At this point, the mixture forms an azeotrope--the concentration at which the ethanol and water vaporize at the same temperature. Unless other methods are used, no further concentration is possible (Stark 1954; Schroder 1981).

The residue remaining in the water separated in the stripping column is called stillage, and consists of the unfermented organic materials from the mash. If the stillage is not used as a coproduct, then it must be handled as a waste. It has a high biological oxygen demand (BOD), up to 40,000 ppm, making disposal difficult. Part of the problem is the sheer amount produced; a 4.5-million-liter plant produces 227,000 L/day of stillage. Anaerobic digestion, activated sludge, and trickling filters--all standard biological treatment methods--may be utilized, but represent a high capital expense (Noyes Development Corporation 1964).

Ethanol Drying. To produce anhydrous--water-free--ethanol, the azeotrope must be broken and the ethanol separated from the remaining water. The most common methods for accomplishing ethanol drying are (1) azeotropic distillation, (2) use of desiccants, and (3) use of molecular sieves (Winston n.d.b; Schroder 1981).

Azeotropic distillation may be accomplished in a reflux distillation column with the addition of a third chemical, such as benzene or n-hexane. The third chemical, or solvent, breaks the azeotrope by changing the boiling characteristics of the mixture and permits distillation to continue until only pure ethanol remains. Most of the ethanol exits through the bottom of the column. Water vapor, solvent, and the remaining ethanol exit through the top of the column and are condensed and then processed through a liquid phase separator, in which they are separated into an ethanol-solvent stream and a water-solvent stream. The ethanol-solvent stream is refluxed back in the first column, and the water-solvent stream is processed in an additional column for solvent recovery (Hawley 1981; Winston n.d.b).

Chemical desiccants used in ethanol drying form a stable compound by reacting with the water, but do not react with the ethanol. One such chemical is calcium oxide (C_aO). The reaction of C_aO and water gives off heat, which must be considered within the system design. Various forms of starch may also be used as desiccants (Ladisch 1979; Winston n.d.b).

Molecular sieves are crystalline aluminosilicates. The sieve material undergoes hydration and dehydration with little or no change in crystal structure. The molecular sieve selectively adsorbs water, because the pores of the crystal are smaller than the ethanol molecules, but larger than the water molecules. The ethanol--in either vapor or liquid form--is passed through one column until the material in that column can no longer adsorb water. Then the flow is switched to the second column, while hot air or gas is passed through the first column to evaporate the water. In some systems, only one column may be used (Winston n.d.b; Schroder 1981).

Ethanol Storage. The primary consideration in anhydrous or dry ethanol storage is to keep out moisture. Fuel tanks with air-tight bladders are effective means for storing dry ethanol. Other considerations include avoidance of ignition sources or fire hazards.

Coproduct Treatment. Coproducts of ethanol production are stillage, carbon dioxide, and fusel oil. During fermentation, the yeast grows rapidly. It is possible to recover this additional yeast and treat it as a coproduct. The yield is .6 lb/gal of ethanol. Additional specialized equipment would be required.

The value of the stillage as a high-protein feed is dependent on the original feedstock used, but may be enhanced by the yeast remaining in the mixture. Grain feedstocks provide the most valuable stillage for use as animal feed. Milo stillage, for example, has a protein content of 31%, while molasses stillage has a protein content of only 6.95% (Schroder 1981; Hodge 1954).

Stillage, or nonfermentable materials, can be removed at any one of four points during production: (1) during the milling process, (2) after cooking and before fermentation, (3) after fermentation and before distillation, and (4) during or after distillation from the bottom of the still (thus the derivation of the name stillage). The first option requires sophisticated and expensive equipment, the second produces a clear wort and permits yeast recycling but fermentable material is lost; the third simplifies distillation but ethanol is lost; the fourth is most commonly used, particularly in small-scale plants and results in the least amount of ethanol loss (Winston n.d.b).

After removal of the stillage, it is separated into solids and liquids (thin stillage). After initial screening and processing through a dewatering press, the solids still contain about 65% water and are called wet stillage. This stillage may be fed directly to livestock. One head of cattle will consume the stillage coproduced with 1 gal/day of ethanol. Storage for any length of time is difficult because of potential microbial contamination. Further drying is necessary if long-distance transportation or lengthy storage is required. Once dry, the stillage must only be protected from insects and rodents.

The thin stillage of fermented grains also contains valuable proteins and carbohydrates, but is difficult to dry further. It is acidic and may be applied directly to the soil through irrigation systems. It may also be anaerobically fermented to produce methane.

The gaseous carbon dioxide is usually purified through adsorption or absorption and may be compressed to liquid form.

Fusel oil is a mixture principally of n-amyl, n-butyl, isobutyl, n-propyl, and isopropyl alcohols, and also contains acids, esters, and aldehydes. It is used as a chemical feedstock, solvent, and fuel. The fusel-oil fraction may be recovered during distillation near the bottom of the column--in the case of a plate column three or four plates above the stream feed point--with a second fraction recovered in the upper column. The lower fraction represents approximately 10% of the product stream, and the upper fraction 5%. It may be stored in fuel tanks (Hawley 1981; Stark 1954).

FEASIBILITY CONSIDERATIONS

Feedstocks

The monomeric sugar substrate used for fermentation ethanol production may be derived from a variety of agricultural sources, such as (1) sugar crops, (2) starch crops, (3) cellulose crops, and (4) agricultural product wastes.

Feedstocks affect not only the amount and characteristics of coproducts, but also the use. For example, when cereal grains are used as the feedstock, a high-protein stillage is coproduced. This stillage has many potential markets. Sugarcane, on the other hand, does not produce a high-protein waste material.

Obtaining fuel from agricultural feedstocks has occasionally been interpreted as obtaining fuel from food. As important as the development of domestic fuel is, adequate food production certainly has a priority. One of the criteria for developing alcohol fuels, or any other form of alternative energy, is that they must not adversely affect essential industries, such as food production. However, it is feasible to obtain the necessary feedstocks for producing fermentation ethanol at a reasonable level without decreasing the human food supply. Several means are available for accomplishing this.

One obvious possibility is to convert to ethanol production a percentage of feed crops (crops used for livestock feed), which includes 90% of the corn grown in the United States (U.S. Department of Agriculture 1978). The advantage is that not only is fuel produced, but a high-protein stillage coproduct is obtained at the same time. The production of ethanol from agricultural feedstocks uses the carbohydrates and leaves the remaining nutrients in a concentrated form. For example, the stillage from grains contains three times the protein in the original feedstock by weight. This coproduct is an excellent animal feed. This stillage lacks the carbohydrates needed in animal feed, but carbohydrates can be added by using readily available and inexpensive low-protein, high-carbohydrate forage crops.

Putting idle land into production can also increase the capacity for growing crops for fuel. Marginal land that could not economically support food crops may sometimes be used productively for fuel feedstocks.

In 1978, the United States produced 500 million bushels of wheat in excess of disappearance. Disappearance is based on domestic production plus imports, minus domestic consumption and exports (U.S. Department of Agriculture 1978). This wheat alone could have produced 1.30 billion gallons of ethanol based on 2.6 gal/bu.

An estimated 5% - 10% of the grain produced annually in the United States is considered distressed, or marginal, and unfit for human consumption. In 1978, 12 billion bushels of grain were grown in the United States (U.S. Department of Agriculture 1978). Using the 5% figure, that equals 600 million bushels of unusable grain. This could have been converted to 1.50 billion gallons of ethanol based on 2.6 gal/bu.

In addition, a certain amount of perishable produce spoils in the fields, during transport, and on the shelf. At this time, the only portion that is practical to collect is that which spoils in the fields. However, it may be possible to establish a collection system for the remainder of the spoiled crops and convert these to alcohol.

Finally, changes in agricultural practices could have great impact on the availability of crops for fuel production. One example is the cultivation in semiarid areas of fodder beets, mesquite, or Jerusalem artichokes, which are potential fermentation ethanol feedstocks. Other possibilities include developing crops specifically for fuel production, and increasing the amount of ethanol that can be produced per acre of feedstock.

Feedstock selection should include the following considerations:

1. production ease and cost of obtaining monomeric sugar;
2. yield of fermentable sugar per acre;
3. yield of residue from the production process per volume;
4. cost, composition, and quality of residue for potential use as a coproduct, such as animal feed; and
5. appropriateness of land-cultivation practices for feedstock cultivation.

The most commonly used feedstock in the United States is corn, followed by milo, potatoes, wheat, barley, sugar beets, molasses, and sugarcane, in that order (Solar Energy Information Data Bank 1981.)

Sugar Crops. Sugar crops include sugarcane, sugar beets, sweet sorghum, Jerusalem artichokes, fodder beets and fruits. Molasses, a by-product of sugarcane refining, may also be considered a sugar feedstock for ethanol production, although it is not classified as a crop. Sugar feedstocks contain individual or bonded pairs of 6-carbon sugars, making processing for preparation of fermentation a simple, mechanical procedure of pressing or crushing, and extracting the juice. Advantages of using sugar crops include low equipment, labor and energy costs; disadvantages include additional considerations necessary for proper storage to avoid contamination.

Sugarcane is grown in a limited geographical area in the United States--Florida, Hawaii, Louisiana, and Texas. Its advantages include (1) high sugar concentration, (2) high sugar yield, and (3) the cane residue, or bagasse, can be used as a fuel for process heat.

The molasses feedstock may be either blackstrap or invert molasses. Blackstrap molasses is a waste product from sugarcane refining. It contains approximately 50% fermentable sugars, primarily as sucrose and invert sugar; approximately 90% of that sugar is fermentable (Hodge 1954). Table 3.5 shows the composition of blackstrap molasses.

Invert or high-test molasses is produced by heating sugarcane juice at acid reaction, neutralizing it, and evaporating it. Approximately 95% of the sugar is fermentable (Hodge 1954).

Sugar beets need to be rotated with nonroot crops on a four-year cycle. They are grown throughout the United States, principally in the Northwest, Southwest, and Midwest. Their advantages include (1) high yield of sugar per acre, (2) high yields of beet pulp and beet top coproducts, and (3) toleration of varying climatic and soil conditions.

Table 3.5
Composition of Blackstrap Molasses

Element	%
Solids	83-85
Sucrose	30-40
Invert Sugar	12-18
Ash	7-10
Organic Nonsugars	20-25
Source: Hodge 1954.	

Jerusalem artichokes are a member of the sunflower family and are native to North America. Grown only on a limited scale for agricultural production, they have definite potential as an alcohol fuel feedstock. Although untested on a commercial scale, their advantages appear to be (1) toleration of varying climatic and soil conditions, (2) toleration of low soil fertility, (3) perennial growth, which means replanting is not required (if stalk rather than tubers are harvested), (4) ease of cultivation, and (5) low susceptibility to diseases and pests.

Fodder beets are a hybrid of sugar beets and mangolds, and are agronomically similar to sugar beets. Their advantages include (1) high yields of sugar per acre, (2) reisistance to loss of fermentable sugars during storage, and (3) lower fertilizer requirements than sugar beets.

Fruit crops are unlikely to be used as a feedstock for alcohol fuel production because of their high market value for direct human consumption. Wastes from fruit processing and distressed fruit are potential feedstocks, however, and are discussed under Agricultural Waste Products.

Starch Crops. Starch crops include the grains-- corn, milo, wheat, and barley--and the tubers--potatoes and sweet potatoes. Starch has a long-branched chain of 6-carbon sugars. It is unlike sugar crops that have individual or paired 6-carbon sugars. The composition of starch presents both an advantage and disadvantage in ethanol production. The advantage is that because sugar is not in a form readily available for fermentation it stores better, with minimal loss of potential fermentable sugars. The disadvantage is that

it is more complicated (thus more expensive) to break
down the chain into monomeric sugar. This is usually
accomplished through enzymatic or acid hydrolysis.
However, when related to total ethanol production costs,
this is a small part of those costs.

Table 3.6 gives a comparison of corn, milo,
potatoes and wheat yields for ethanol production. Table
3.7 gives the basic composition of corn, milo, barley,
and wheat.

Table 3.6
Comparison of Selected Starch Crop Yields

Material	Av. Yield 1977-78 Normal Unit of Sale/Acre[1]	Yield Gallons EtOH/ Normal Unit of Sale	Yield Gallons EtOH/Acre
Corn	95.99 bu/acre	2.6 gal/bu[2]	249.57 gal/acre
Milo	55.7 bu/acre	2.5 gal/bu[3]	139.25 gal/acre
Potatoes	12.5 tons/acre	29 gal/ton[2]	362.50 gal/acre
Wheat	31.1 bu/acre	2.6 gal/bu[2]	80.86 gal/acre

Source: 1. U.S. Department of Agriculture 1979.
2. University of Nebraska 1980.
3. Hedrick 1980.

Cellulose Crops. Commercialization of alcohol
production from cellulose crops has not been
accomplished in the United States. However, this is a
major emphasis of current research and development
because of the low cost and high availability of
cellulose as a potential feedstock for alcohol fuels.
There are plans for developing experimental large-scale
pilot plants utilizing cellulose. The extended chains
permitted by the β-glycosidic linkages in cellulose
create increased hydrogen bonding over starches. This
results in lower solubility, less tendency to swell, and
less accessibility to the reagent (Saeman 1954).

Agricultural and Forest Product Wastes. These
wastes include food-processing waste, such as cheese
whey and cannery waste, including both vegetable and
fruit wastes; wood waste; and lignosulphonate waste from
the sulphite process used in certain pulp mills.

Cannery waste can be processed to provide
single-cell protein, methane, or ethanol. Ethanol
production from cannery waste remains commercially
unproven.

Cheese whey is easily fermentable and contains 6.5%
solids, including 4.5% - 5% sugar in the form of
lactose, 0.8% protein, and mineral salts. Efficient

Table 3.7
Composition of Various Grains

	Corn	Milo	Barley	Wheat
Digestible energy Kcal/kg[1]	3,610.0	3,453.0	3,080.0	3,520.0
% Protein[1]	8.9	11.0	11.6	12.7
% Lysine[1]	.18	.27	.53	.45
% Methionine - cystine[1]	.18	.27	.36	.36
% Tryptophan[1]	.09	.09	.18	.18
% Calcium[1]	.02	.04	.08	.05
% Phosphorus[1]	.31	.29	.42	.36
% Fiber[1]	2.0	2.0	5.0	3.0
% Ether Extract[1]	3.9	2.8	1.9	1.7
Carbohydrates gm/lb[2]	55.1	448.8	315.2	170.4
	(Kernels)			(Flour)

Source: 1. Texas Department of Agriculture 1978.
         2. U.S. Department of Agriculture 1975.

ethanol production from cheese whey would require each dairy processing plant to have its own production plant on-site. Initial calculations indicate that transportation of the whey more than 100 miles would make costs prohibitive. The average dairy plant yields 30,600 lb/day of cheese whey, which in turn provide 1,200 lb/day of lactose that could be converted into about 91 gallons of ethanol. In 1980 the Milbrew Company in Juneau, WI produced 2 million gal/yr of 193-proof ethanol from cheese whey (Bailey 1980).

One problem with cheese whey is the high biological oxygen demand (BOD) of the residue of the ethanol production process. If a sewage treatment type of facility is not used, it becomes a major pollutant.

Wood wastes share the problem of fermentation of cellulose crops--the difficulty in hydrolyzing cellulose. Strong acid or dilute acid may be used for the hydrolysis of wood wastes. Approximately 80% of the sugars may be fermented. The lignin residue equals approximately 30% of the weight of the wood; approximately 11 pounds of lignin are released for every gallon of alcohol produced. Lignin may be used as a fuel for process heat in the plant (Saeman 1954).

Ethanol fermentation plants using wood as a feedstock have been operated in Switzerland and Germany and in the United States by what was Vulcan Copper & Supply in Cincinnati, Ohio. The high processing costs necessitate an efficient, large-volume facility.

The sulphite pulping process was introduced in the mid-1860s. It consists of treating wood chips at a high temperature with an aqueous solution of sulphuric acid and a reactant, such as calcium, magnesium, sodium, or ammonium. The lignin and hemicellulose are removed from the fibrous cellulose, which is used to produce paper and other products. The remaining sulphite liquor contains organic substances equal to approximately 50% of the weight of the raw wood and the chemicals used in the reaction process. The sugars contained in sulphite waste liquor include: glucose, mannose, galactose, fructose, xylose, and arabinose. The first plants utilizing this residue for fermentation ethanol were built in Sweden in 1907. Commercial activity centers in Sweden and Germany (McCarthy 1954).

Of the 11 sulphite mills in the United States, only the Georgia Pacific Co. has an alcohol-production plant connected to a pulp mill. Located in Bellingham, WA, this plant has been operating since 1944. The plant was originally built to produce ethanol for use in manufacturing synthetic rubber during World War II. In 1980, it produced 5.3 million gallons of 190-proof ethanol, of which 1.8 million gallons were converted to 200-proof ethanol.

The Georgia Pacific operation is commercially viable for the following reasons:
1. the plant was built in 1944, so today's capital costs are not a factor;
2. a continuous fermentation process is used, maximizing production; and
3. the company has a strong research program on use and market development of desugared lignin, the waste from the fermentation process.

This lignin can be used in many industrial processes; for example, as a resin or as a filler. Georgia Pacific holds several patents on lignin production and use, giving the company an advantage in the marketplace.

Alcohol fermentation plants located on-site with the pulp mill would be essential. Most pulp mills operate 350 days a year and could produce 26 gallons of 190-proof ethanol/ton of pulp. The potential alcohol production from these pulp mills is slightly more than 17 million gal/yr of ethanol. Table 3.8 lists those plants using the sulphite process to produce pulp; most of them, however, burn the wastes for process heat, which at this time is a more cost-effective use for the waste.

Table 3.8
Sulphite Process Pulp Mills

Pulp Plant; Location	Tons Pulp/Day
ITT - Rayonier - Hoquiam WA	450
Finch - Pruyn, NY	200
Crown Z - Lebanon, OR	100
Consolidated Paper - Appleton, WI	130
Boise-Cascade - Salem, OR	200
Wausau - Wausau, WI	160
American Can - Green Bay and Rothschild, WI	350
Flambeau - WI	100
St. Regis - WI	80
Badger Paper - WI	100
Source: Georgia-Pacific Co. 1980.	

Coproducts

The coproducts in fermentation ethanol production are stillage, carbon dioxide, and fusel oils. The composition of the stillage and the fusel oils is a function of the feedstock. The yeast used for fermentation grows rapidly during the process and may be recovered, although few commercial operations follow this procedure. It is 50% protein, and the yield is.6 gal/gal ethanol.

Starch Crop Stillage. The stillage from starch crop fermentation contains nonfermentable material in the form of solids in suspension, ranging from very fine to very coarse, and water-soluble material in solution. The stillage can be directly applied to land as a fertilizer, but the amounts must be carefully monitored because of its acidity and odor. The coarser solids may be separated out, and when grain is the feedstock, these are called distillers grains. When dried and blended with the dried soluble material, these solids are called distillers dark grains (DDG) or distillers dried grains with solubles (DDGS). (The acronym "DDG" is used for both distillers dark grains and distillers dried grains; DDGS is used here to represent the combination of dried grains and solubles, and DDG is used to refer to distillers dried grains only.) The nutritional content of these products will vary according to the feedstock and production process used.

All the stillage products from beverage distilleries have been used for some time as animal feed in areas near these distilleries. The bourbon industry is the primary source, with corn the original feedstock. Stillage from wheat and rye are also used. In addition to the value of distillers feeds for protein and energy, research in the 1950s showed such feeds stimulated certain digestive processes in ruminants (cud-chewing animals), primarily in cellulose digestion (Distillers Feed Research Council n.d.). The uses of distillers feeds for animals includes:

Poultry - Layers and Breeders: Distillers feeds contain protein, fats, known and unknown vitamins and minerals, and are a rich source of linoleic acid. They provide excellent poultry production and hatchability. Feeding formulas using up to 20% distillers dried grains with solubles (DDGS) have been shown to produce very satisfactory results.

Poultry - Broilers and Chick Starters: The high vitamin content of distillers dried solubles (DDS) makes it an excellent ingredient in chick starter and broiler rations. Eighty percent of the corn in a poultry diet can be replaced by DDS from yeast fermentation of cereal grains, and can be mixed in an all-mash ration in the proportion of two parts syrup to one part mash. DDGS also may be used, and is an excellent source of riboflavin as well as vitamins.

Turkeys: DDGS can be used to levels up to 10% in turkey diets. Positive effects on egg production and hatchability have been reported.

Dairy Cattle: Distillers feeds, when used as a

silage additive, have a positive effect on milk production, and for the lactating cow they are palatable and highly digestible. Studies also indicate a favorable effect on milk percentage.

Calves - Starters: DDS and DDGS can replace ingredients such as linseed meal and dried whey in starter formulas for calves.

Sheep: Little research has been done on the effect of using distillers feeds as supplements for sheep. The research that has been carried out indicates characteristics of palatability, digestibility, and the stimulating effect on rumen function, making DDGS a good feed supplement.

Swine: DDGS can be used in swine diets up to 10%. Several studies have been conducted using various distillers feeds in swine diets. The consensus is that they provide an excellent source of feed.

Dogs: Both growth and reproductive studies with dogs indicate that DDS is an excellent dog food supplement.

Fish: Studies indicate that distillers feeds can be used in formulas for both trout and warm-water fish, such as catfish. In the trout diet approximately 21% DDS was used.

Mink: Studies using DDS to replace more expensive feed ingredients in mink diets have produced growth and pelt quality equal to traditional diets.

Beef Cattle: Distillers feeds provide a palatable and highly digestible protein source for beef cattle. They also help utilize cellulose and urea, and stimulate growth when used in liquid feed supplements as a source of the urea-protein factor (Distillers Feed Research Council n.d.).

Sugar Crop Stillage. The residue from molasses fermentation has been used as an animal feed, primarily for dairy cattle; as an antidusting agent in feed mixing and handling; and as a fertilizer. Table 3.9 shows the analysis of molasses residue evaporated to 45% - 50% solids.

Residue from molasses fermentation is useful as a fertilizer because it contains potash salts, nitrogen compounds, and phosphates. It can be applied directly to the soil as a liquid, but costs and logistics would require use close to the fermentation plant. Excessive use can create problems, such as a proliferation of insect pests.

Carbon Dioxide. The primary uses of carbon dioxide (CO_2) are in the manufacture of carbonated beverages, fire extinguishers, dry-ice production, food processing,

Table 3.9
Analysis of Molasses Residue Evaporated to 45% - 50%
Solids

ANALYSIS, WET BASIS	%
Moisture	54.67
Solids	45.33
Protein	6.95
Ash	10.93
Gums	10.40
Sugars (copper-reducing substances)	5.30
Glycerol	2.60
Lactic acid	2.70
Fiber	0.30
Wax, lignin, glucosides, phenolic bodies, organic acids, etc.	6.15
ASH ANALYSIS, WET BASIS	%
Silica	0.4
Iron	0.08
Aluminum	0.09
Calcium	1.4
Magnesium	0.7
Sulphur	1.4
Phosphorus	0.3
Sodium	0.5
Potassium	3.4
Chlorine	1.3
Manganese	0.002
Iodine	0.0014
Copper	0.017
VITAMIN ANALYSIS, WET BASIS	mg/lb
Nicotinic acid	9.53
Pyridoxine	13.62
Pantothenic acid	17.71
Biotin	0.68
Folic acid	0.135
Riboflavin	3.63

Source: Hodge 1954.

and the chemical industry. Approximately 70% - 80% of the CO_2 produced in the fermentation ethanol process can be recovered. In the fermentation of molasses, for every 100 pounds of feedstock fermented, approximately 44 pounds of CO_2 are produced; therefore, 30.8 - 35.2 pounds of CO_2 are recoverable (Hodge 1954).

Uncleaned, uncompressed CO_2 may be sold, but usually the gaseous CO_2 is purified after it is collected. Although it contains relatively few impurities--approximately 0.5% solids, aldehydes, and alcohol--they cause an odor. These impurities may be purged by adsorption or absorption.

Gaseous CO_2 may be converted to liquid CO_2 in compressors. Liquid CO_2 may be converted to solid CO_2 by rapid evaporation and compression.

Fusel Oil. Fusel oil is a combination of higher alcohols. For example, the fusel oil from molasses fermentation comprises isopropyl, n-propyl, isobutyl, n-butyl, isoamyl, and d-amyl alcohols (Hodge 1954).

Fusel oil is used as a chemical feedstock and as a solvent. It also may be burned as fuel. In large-scale fermentation plants, it could be removed in an extra action column, fractioned, and sold as a coproduct. This is out of the range of small-scale production, where fusel oil merely adds to the heating value of the ethanol produced. [Molasses fermentation, for example, generally yields about 1.1 L fusel oil/1,000 kg feedstock (Paturau 1969); with corn as the feedstock, fusel oil comprises approximately 3% of the alcohol produced.]

Economics

The criteria for determining feasibility includes (1) investment and operating costs, (2) production factors, (3) market potential, and (4) energy analysis.

Investment and Operating Costs. The feasibility study conducted for any particular project should indicate the economic factors influencing the size of the facility. Whatever the facility size, the production process will remain essentially the same. A regional approach must be taken to energy development, production, and distribution. Such an approach takes into consideration feedstock availability, potential markets, and volume of those markets. Small-scale (<15 million gal/yr) on-farm production provides an alternative use of crops, an additional "cash crop," and increased self-reliance. Large-scale production (>15 million gal/yr) provides ethanol for blending with gasoline for widespread distribution or for use as a chemical feedstock and solvent.

Investment costs include:
1. land;
2. buildings;
3. plant equipment;
4. initial feedstock inventory;
5. initial yeast and enzyme supplies;
6. taxes;
7. insurance;
8. depreciation;
9. interest on loan or mortgage;
10. business formation;
11. licensing costs; and
12. equipment installation.

Operating costs include:
1. labor;
2. maintenance;
3. taxes;
4. supplies;
6. delivery
7. energy;
8. water;
9. insurance;
10. interest on debt; and
11. bonding.

For large-scale ethanol production plants, the income derived from the sale of products is directly compared with the estimated capital and operating costs to determine the project's economic feasibility. It is essential to collect the most recent information applicable to a specific project's production capabilities, product costs, and market potential.

In small-scale plants, the market value of products may depend to a significant degree upon local supply and demand. Therefore, direct assessments of current local market conditions may be used to determine direct income from sales. It is essential that local markets be evaluated realistically when determining plant capacity. For products used directly, on-site costs may be determined by comparing them to the least-expensive purchased alternative. This is true for both alcohol fuel and any stillage used as animal feed. In addition, small integrated (crop and animal producing) farms receive other benefits, such as energy independence for farming operations and an alternative market for farm commodities. While these are difficult to quantify, they may be included as factors in evaluating income and benefits.

Production Factors. Production factors can be categorized as follows:
1. use;
2. scale;
3. feedstock availability;

4.   yield;
5.   resource requirements;
6.   labor requirements;
7.   plant design; and
8.   plant operation.
The end use of the ethanol produced is a major factor in plant design and production economics. If anhydrous (200-proof) ethanol is required, the ethanol drying step must be included in the production process. These are options, however. In small-scale production, ethanol drying may take place at a centrally located blending site, rather than at the fermentation facility. Initial determination of ethanol proof to be produced is essential in determining feasibility. In addition, recovery and use of the coproducts must be evaluated.

Fermentation ethanol production is appropriate in both small- and large-scale oprations. Small-scale fermentation ethanol production may be considered a means for expanding farm operations and developing an alternative market for excess or increased production. A regional cooperative venture could provide participants a continuing, stable source of fuel for farm vehicles.

Large-scale fermentation ethanol production can provide the United States a direct savings on crude oil imported through the addition of ethanol to gasoline to produce E10, commonly called Gasohol (a blend of 10% agriculturally derived anhydrous ethanol and 90% unleaded gasoline) or other gasoline-ethanol blends. Such blends decrease the amount and cost of crude oil imported not only on the basis of the 10% petroleum displaced per gallon by ethanol. In the past, the octane of fuels was increased by adding tetraethyl lead. Because the lead compounds have significant adverse impacts on the environment, the conversion to unleaded gasoline was mandated by the U.S. Environmental Protection Agency. The changes in refinery operations required to produce fuel of the same octane without lead reduce the quantity of fuel that can be produced from a barrel of crude oil. This results from altering the chemical constituency of the gasoline by reforming lower hydrocarbons to increase the percentage of octane-boosting aromatic compounds. This reforming process consumes additional energy in the refining process-- energy directly lost from every barrel of crude oil. The addition of ethanol to gasoline effectively gives the required octane boost, and the reforming requirement is correspondingly reduced (Jawetz 1979).

Feedstock availability includes (1) feedstock(s) selection, (2) cost (historical, current, projected), (3) continuing availability in required amounts, (4) cost of collection and transportation, and (5) contingencies.

Theoretical fermentation production yields are generally 95% of the potential possible chemically (the other 5% is allocated to yeast-cell production and by-product production). Fermentation efficiency is a ratio between actual ethanol produced and the theoretical yield.

The primary resource requirements are water and process fuel. Approximately 16 gallons of water are needed for every gallon of ethanol produced. Biomass may be used as a process fuel, as well as the more traditional coal and natural gas. Table 3.10 shows process fuel sources used in fermentation ethanol plants operating in the United States as of March 1981 with their frequency of use by gallons of ethanol produced. Additional potential process fuels for which operational data does not exist include bagasse, methane, syngas, and waste heat. Table 3.11 lists a selection of process heat sources and factors affecting their use.

Table 3.10
Frequency of Process Fuel Use

Process Fuel	Frequency by Gal/Ethanol Capacity
Alcohol	150,000
Biomass	302,400
Butane	400
Coal	1,096,000

Source: Solar Energy Information Data Bank 1981.

Labor required for operation of a fermentation ethanol plant is low and depends on the size of the plant, amount of automatic process-control equipment used, and time in operation. Adequate training is a prerequisite.

Plant design is primarily affected by size and site considerations. Size depends on feedstock availability and market potential. Site selection is based on the following factors:
1. climate,
2. topography,
3. subsurface land strata,
4. water availability,
5. transportation considerations,

Table 3.11

Process Heat Source and Factors Affecting Use

Heat Source	Heating Value (dry basis)	Form	Special Equipment Required	Boiler Types	Source	Advantages	Disadvantages
Agriculture Residues	3,000-8,000 Btu/lb	Solid	Handling and feeding eqpmt.; collection eqpmt.	Batch burner-fire tube; fluidized bed	Farm	Inexpensive; produced on-farm	Low bulk density; requires very large storage area
Coal	9,000-12,000 Btu/lb	Solid	High sulfur coal requires stack scrubber	Conventional grate-fire tube; fluidized bed	Mines	Widely available demonstrated technology for combustion	Potentially expensive; no assured availability; pollution problems
Waste Wood	5,000-12,000 Btu/lb	Solid	Chipper or log feeder	Conventional fluidized bed	Forests	Clean burning inexpensive where available	Not uniformly available
Municipal Solid Waste	8,000 Btu/lb	Solid	Sorting eqpmt.	Fluidized bed or conventional fire tube	Cities	Inexpensive	Not widely available in rural areas
Pyrolysis Gas		Gas	Pyrolyzer-fluidized bed	Conventional gas boiler	Carbonaceous materials	Can use conventional gas boiler	Requires additional piece of equipment

Table 3.11, continued

Heat Source	Heating Value (dry basis)	Form	Special Equipment Required	Boiler Types	Source	Advantages	Disadvantages
Geo-thermal	N/A	Steam/hot	Heat exchanger	Heat ex-changer water tube	Geo-thermal	Fuel cost is zero	Capital costs for well and heat ex-changer can be extremely high
Solar	N/A	Radi-ation	Collectors, concentra-tors, stor-age bat-teries, or systems	Water tube	Sun	Fuel cost is zero	Capital costs can be high for required equipment
Wind	N/A	Kine-tic energy	Turbines, storage batteries, or systems	Elec-tric	In-direct solar	Fuel cost is zero	Capital costs can be high for required equipment

Source: Solar Energy Research Institute 1980.

6.   integration with related industries, e.g., use of excess process heat or a feedlot for direct stillage use, and
7.   ground water.

Specific plant design is a result of unit process selection and integration. The unit processes comprise (1) feedstock storage, (2) feedstock preparation, (3) fermentation, (4) distillation, (5) ethanol drying (for anhydrous ethanol), (6) ethanol storage, and (7) coproduct treatment and storage.

Plant operations include (1) management, (2) receiving and shipping, (3) feedstock preparation, (4) fermentation, (5) distillation, (6) enzyme and yeast culturing, (7) denaturing, (8) maintenance, and (9) quality control.

Market Potential. Ethanol has three primary uses (1) as an engine fuel, (2) as an industrial chemical, and (3) as a solvent.

As an engine fuel, ethanol may be used in the following ways:
1.   ethanol-gasoline blends;
2.   hydrated (lower-proof) ethanol;
3.   straight (neat) anhydrous (200-proof) ethanol; and
4.   dual-carbureted diesel fuel supplement.

Ethanol may be used in spark ignition engines, turbine engines, and diesel engines (when dual carbureted).

The chemical industry uses ethanol as a feedstock and a solvent. Ethanol for industrial uses is primarily produced from ethylene, a by-product of petroleum refining. Ethylene is used widely in the production of polymers. While ethanol is produced from ethylene, it may also be dehydrated to produce ethylene. Ethanol can be fermented to acetic acid, which is widely used in polymer production. For use as a solvent, ethanol must be anhydrous and free of all impurities.

The market already exists for Gasohol and other ethanol-gasoline blends. In 1979, 50 million gallons of ethanol were blended with gasoline for sale as a motor vehicle fuel. No engine modifications are necessary and public acceptance for this clean, high-octane fuel continues to increase.

The use of hydrated ethanol requires certain engine modifications. Therefore, the most practical use is in captive fleets, e.g., a specific city's taxis or municipal vehicles.

The cost of producing anhydrous ethanol is high; therefore, it is unlikely to be used as a straight fuel.

While ethanol can be used as a dual-carbureted diesel fuel, the liabilities appear to outweigh the benefits. Butanol, derived from the same feedstocks as ethanol, may have potential as a diesel fuel.

Energy Analysis. Considerations for evaluating the energy balance of producing fermentation ethanol for use as a liquid fuel should include the following:

1. energy requirements for feedstock production;
2. energy requirements for fermentation ethanol production;
3. system efficiency, particularly heat transfer potential; and
4. liquid fuel energy balance.

There are three primary areas in which mistakes are made in determining energy balances: (1) attributing unnecessary costs to the production process, (2) attributing coproduct treatment costs to the cost of ethanol production, and (3) ignoring the liquid fuel energy balance.

Unnecessary costs that may be attributed to the production of corn as a feedstock for ethanol production, for example, include the costs of irrigation or electricity. The majority of the corn--96%--grown in the United States is dry land corn, thus irrigation costs are irrelevant, as are electricity costs since electricity is not used to grow crops.

Any energy costs related to coproduct treatment, such as stillage drying or carbon dioxide recovery, should not be included in the energy costs of ethanol production.

Finally, the use of the product must be taken into consideration. If electricity is needed, for example, 10 Btus of coal are burned in a coal-fired plant to produce 3.3 Btus of electrical energy. The energy cost is acceptable because electricity has so many advantages. It can be transported great distances; its transmission systems occupy relatively little space; its uses are varied; it can be used in unique ways, e.g., electrical lighting. Similarly, if liquid fuel is needed for transportation, the energy costs must be evaluated in terms of net liquid fuel gain.

The exact energy requirement for producing any feestock is dependent upon the region, growing season, soil characteristics, availability of water, and cultivation requirements for the crop. Therefore, no general analysis can define a precise energy requirement. However, consideration of the representative data does establish approximate requirements, and these are valid as long as the limitations are observed.

Corn will be used to set up an energy-cost example, since corn is the major feedstock for ethanol production in the United States, and since data exists for energy requirements for corn production. Of the corn produced in the United States, only 4% needs to be irrigated, therefore, this cost will not be included. Energy

expenditures directly attributable to corn production
are (1) machinery, (2) fuel, (3) fertilizer, (4)
insecticides and herbicides, and (5) drying.

Agricultural Engineers Handbook gives an estimated
dry land corn machinery energy cost of 373,000 Btu/acre.
This estimate takes into account the energy requirements
for equipment manufacturing and equipment life (Wittmuss
1975).

The fuel requirement covers fuel expended for
plowing; discing; planting; spreading fertilizers,
herbicides, and pesticides; cultivating; and harvesting.
The energy cost is estimated to be 5.15 gal/acre of
diesel fuel. Assuming a diesel fuel calorific value of
140,000 Btu/gal, this equals 721,000 Btu/acre (Wittmuss
1975). The energy cost of fuel to transport grain to
storage--assuming 448 bu/load, an average transport
speed of 15 mph, and a distance of five miles to
the farmstead--is an estimated 1.81 gal/production acre
of diesel fuel, or 253,400 Btu/acre.

Because repeated plantings of corn deplete the
soil, extensive use of nitrogen and phosphate
fertilizers are used to maintain production rates. An
estimated 125 pounds of nitrogen and 20 pounds of
phosphate are used per acre (Wittmuss 1975). The
energy expended in production of these fertilizers
amounts to approximately 33,331 Btu/lb each (Pimentel
1979). This means an energy cost of 4.16 million
Btu/acre for nitrogen and 666,000 Btu/acre for
phosphate. Although corn can be grown without using
insecticides and herbicides, they are used on many
farms. An estimated 1.3 pounds of insecticides and 2.5
pounds of herbicides are used per acre (Rosell 1974).
Both require 43,648 Btu/lb to produce. Therefore, the
energy cost of corn production is 56,742 Btu/acre for
insecticide and 109,120 Btu/acre for herbicide. The
energy costs of dry land corn production are summarized
in Table 3.12.

Although corn for ethanol production does not need
to be as dry as that for indefinite storage, it must be
dry enough to prevent spontaneous combustion. Assuming
that the corn is dried from 23% moisture to 15%
moisture, 1.2 million Btu/acre could be required
(Winston 1980).

Assuming dry land corn production conservatively
averages 75 bu/acre, the per-bushel energy cost for
production would be 100,680 Btu. Assuming production of
2.5 gallons of ethanol per bushel of corn, this amounts
to an agricultural energy expenditure for ethanol
feedstock production of 40,272 Btu/gal. It should be
noted that some successful farming operations do not use
large amounts of nitrogen fertilizer, phosphates,
herbicides, or insecticides. As the cost of these
chemicals increases in terms of money and energy,

farmers may look to alternative methods (Lockeretz 1980; Keppler 1977). This would reduce the rate of Btu/acre for crop production.

Table 3.12
Summary of Dry Land Corn Production Energy Costs

Expenditure	Btu/Acre
Machinery	373,000
Fuel (Cultivation)	721,000
Fuel (Transportation)	253,400
Fertilizer	4,826,000
Insecticides & Herbicides	100,390
Drying	1,206,000
TOTAL	7,551,000

The energy use in feedstock conversion to ethanol is primarily for cooking (during liquefaction) and distillation.

The energy requirement for cooking can be estimated by assuming a batch fermentation and production of a 10% ethanol beer. Using a starch-heat capacity of 0.3 Btu/lb/°C, an average ambient temperature of 21°C (69.8°F), and assuming a 10% average heat loss, the energy cost would be 10,000 Btu/gal. This calculation is based on mash sterilization by boiling and does not give credit for heat recovery (Winston 1980).

The stillage is a high-protein feed source, and it is unlikely that it would ever be treated as a waste product. If it is, however, the energy cost for drying to lower its BOD to meet environmental standards, must be included. If the stillage is used as a coproduct, then any preparation costs are attributed to that process, rather than to ethanol production. Table 3.13 gives the quantities of energy expended in ethanol production.

Table 3.13
Energy Used for Ethanol Production

Process	Btu/gal
Feedstock Production	40,272
Conversion to Ethanol	30,000
TOTAL	70,272

The total energy needed to produce ethanol fuel depends upon how efficiently the system uses the heat input. Some heat losses are inevitable, but in many cases the overall thermal efficiency can be improved by heat exchange between hot and cold liquids during the process.

The unavoidable heat losses are:

1. the heat loss resulting from inherent inefficiencies in the boiler;
2. the heat loss from piping in and transporting of steam from the boiler to the equipment in which it is used;
3. the inefficiencies and inherent losses in heat-exchange equipment;
4. heat losses in the vessels; and
5. low-grade heat that cannot be effectively used within the process.

The following process steps can conserve energy:

1. The temperature differential after sterilization and before sacchaarification can be exploited to preheat or heat up other process streams or liquids; before a mash can be saccharified, it must be cooled down from 93.3°C (200°F) to less than 60°C (140°F).
2. The heat in stillage upon discharge from the beer still can be run countercurrent to the liquid being introduced into the still in a heat exchanger. The stillage leaving the beer still has the highest temperature and greatest volume of all the liquids in the entire process, and the liquid being introduced into the still has to be heated up from room temperature to the operating temperature of the still.
3. The heat of condensation can be used for process preheat. Every pound of alcohol vapor that comes off the top of the rectification column has to be condensed to be recovered. This requires about 400 Btu/lb.

Ethanol production requires liquid fuel only in the form of diesel fuel and gasoline consumed in crop production and transportation. The total liquid fuel requirements for crop production and transportation for dry land corn amount to 1 gal/27 gal ethanol produced. (The liquid fuel energy balance is discussed in more detail in Chapter 1.) All other energy needs can be provided by nonpetroleum energy sources, such as coal, biomass, and where applicable, geothermal energy. Consideration should be given to the use of renewable energy resources where possible.

Environmental and Safety Considerations

Environment. With proper precautions, no major environmental or health hazards arise from fermentation ethanol production. Areas of concern are (1) removal of crop residue from the land for use as a boiler fuel, and (2) use of stillage on the land. Crop residues are important because they help control soil erosion through their cover and provide nutrients, minerals, and fibrous material that help maintain soil quality. However, not more than one-third to one-half of the residues from a grain crop devoted to ethanol production need to be used to fuel the process. Also, several methods for lessening the impact of crop residue removal--such as crop rotation and winter cover crops--can be used.

Two kinds of problems can result from applying thin stillage to the land: odor and acidity. The effects can be attenuated by using a sludge plow, recycling the thin stillage within the plant, and using anaerobic digestion to reduce the pollution potential (Winston 1980).

Environmental and health hazards that could result from increased use of ethanol as a motor vehicle fuel include emissions and spills. Most regulated emissions are lower for ethanol than for gasoline. Aldehydes, an unregulated emission that may be potentially hazardous, is higher. Neither aquatic nor land-based spills would be as serious as gasoline spills because of lesser toxicity and miscibility in water (Moriarity 1981).

Safety. Etahnol is categorized as a Class IIB flammable liquid, which includes those substances with a flash point below 22.8°C (73°F) and a boiling point at or above 37.8°C (100°F). Ethanol's specific flash point is 12.7°C (54.9°F) and its boiling point is 78.3°C (172.9°F). The implications of ethanol's Class IIB status are that certain precautions must be observed when producing or storing it. Although ethanol is less volatile and less potentially explosive than gasoline, in certain mixtures with air it can be ignited by a spark. Therefore, fire prevention must be an integral part of plant design and operations. Precautions include:

1. adequate ventilation;
2. banning open flames within the immediate vicinity;
3. adequate coverage by fire extinguishers or a sprinkling system;
4. adequate and well marked exit routes;
5. elimination of spark sources; and
6. proper equipment maintenance.

In the course of production, precautions can be taken in relation to specific hazards. These are described in Table 3.15.

Dry chemical or CO_2 BC-rated fire extinguishers are adequate for small-scale plants; larger plants require a sprinkler system. Because ethanol and water are miscible, water can be used to extinguish ethanol fires.

Care must be taken with the operation of the boiler providing process heat. Proper maintenance according to design specifications should be adequate to prevent problems.

Below-ground bulk storage of alcohol fuels is required in many areas. In addition, gravity flow of flammable liquids may be restricted in certain urban areas. On-farm storage of ethanol from small-scale plants may not need to meet such stringent requirements, but should be planned with safety considerations in mind.

Large blending sites require a specific mixing room with adequate ventilation, approved explosion-proof wiring, a fire-protection system, and provisions for containment in case of leakage.

The following codes can provide safety guidance and should be considered when constructing an ethanol production plant: flammable liquids code, electrical code, and the life-safety code. Additional codes and standards applicable to ethanol plant design and operation are:

1. Standard for Storage of Flammable and Combustible Liquids on Farms and Isolated Construction Projects—National Fire Protection Association (NFPA) 395;
2. Prevention of Dust Explosions in Industrial Plants—NFPA 321;
3. Basic Classification of Flammable and Combustible Liquids—NFPA 77;
4. Static Electricity—NFPA 30;
5. Flammable and Combustible Liquids Code—Occupational Safety and Health Administration (OSHA) 190.94;
6. Occupational Noise Exposure—OSHA Subpart O;
7. Machinery and Machinery Guarding—American National Safety Institute (ANSI) B31.1;
8. Power Piping—ANSI B31.3;
9. Chemical Plant and Petroleum Piping—Underwriters Laboratory 142;
10. Standard for Steel Above-Ground Tanks for Flammable and Combustible Liquids—American Society of Mechanical Engineers (ASME) Boiler and Pressure-Vessel Code, Section IV and VIII;
11. National Electrical Code—NFPA 70-1978, where applicable (U.S. Department of Energy 1980).

The Occupational Safety and Health Administration of the Department of Labor, in General Industry Standards, revised 7 November 1978, Section 1910.106, provides detailed information on safety standards and equipment specifications for production and storage of flammable liquids (U.S. Department of Labor 1978).

While certain small-scale operations may be exempt
from compliance with these standards, it is important to
review them before plant design is completed to ensure
that the best available safety precautions are
integrated into the design. In addition, operational
procedures should be checked to verify that they are
being carried out in the safest possible manner.

Table 3.14

Hazards and Precautions for Ethanol Production

Hazards	Precautions
Overpressurization; explosion of boiler	Regularly maintained boiler valves that open when pressure exceeds the maximum safe pressure of the boiler or delivery lines.
	Strict adherence to boiler manufacturer's operating procedures.
	If boiler pressure exceeds 20 psi, acquire ASME boiler-operator certification. Continuous operator attendance during boiler operation.
Scalding from steam; steam-gasket leaks	Place baffles around flanges to direct steam jets away from operating areas.
Contact burns from steam lines	Insulate all steam-delivery lines. If electric pump motors are utilized, use fully enclosed explosion-proof motors.
Ignition of ethanol leaks, fumes, or grain dust	Never use metal grinders, cutting torches, welders, or similar tools around systems or equipment containing ethanol. Flush and vent all vessels prior to performing any of these operations.
	Fully ground all equipment to prevent static electricity buildup.
	Never smoke or strike matches around ethanol tanks, dehydration section, distillation columns, or condenser.

Table 3.14, continued

Hazards	Precautions
	(Option) Use hydraulic pump drives; main hydraulic pump and reservoir should be physically isolated from ethanol tanks, dehydration section, distillation columns, and condenser.
Handling acids/bases	Never breathe the fumes of concentrated acids or bases.
	Never store concentrated acids in carbon-steel containers.
	Mix or dilute acids and bases slowly--allow heat of mixing to dissipate.
	Immediately flush skin exposed to acid or base with copious quantities of water. Wear goggles whenever handling concentrated acids or bases; if contact occurs flush eyes with water and immediately call physician.
	Do not store acids or bases over work areas or equipment. Do not carry acids or bases in open buckets.
	Select proper construction materials for components that come into contact with acids or bases, e.g., storage containers, delivery aids, valves.
Suffocation	Never enter the fermentors, beer well, or stillage tank unless they are properly vented.

Source: Solar Energy Research Institute 1980.

REFERENCES

Agricultural Alcohol Blended Fuel Study Conference. 1975. Indianapolis, IN: J.V. Longrock.

Bailey, Richard. 1980. Solar Energy Research Institute, Golden, CO. Personal communication

Bullard, Ron. 1981. Western Energy, Golden, CO. Personal communication.

Casida, F. 1968. Industrial Microbiology. New York, NY: John Wiley and Sons.

Distillers Feed Research Council. n.d. Distillers Feeds. Cincinnati, OH: DFRC.

Draut, Jan. 1980. Georgia Pacific Co., Bellingham, WA. Personal communication.

Flowers, Carey W., Flowers, William J. n.d. "Ethyl Alcohol, Grain Alcohol: Chemistry." Thibodaux, LA: Nicholls State University.

Hawley, Gessner G., ed. 1981. The Condensed Chemical Dictionary, 10th ed. New York: NY; Van Nostrand Reinhold Co.

Hodge, H.M., Hildebrandt, F.M. 1954. "Alcoholic Fermentation of Molasses." In Industrial Fermentations. Underkofler, Leland A., Hickey, Richard J., eds. Vol. 1. New York, NY: Chemical Publishing Co., Inc.

Jawetz, Pincas. 1979. "Alcohol Additivives to Gasoline-An Economic Way for Extending Supplies of Fuel and for Increasing Octane Ratings." Presented at American Chemical Society National Meeting; 9-14 September 1979. Vol. 24, no.3, pp. 798-800. Washington, D.C.: American Chemical Society, Division of Petroleum Chemists.

Keppler, Robert, et al. 1977. "Economic Performance and Energy Intensiveness on Organic and Conventional Farms in the Corn Belt: A Preliminary Comparison."

American Journal of Agricultural Economics. Vol. 59, no. 1, February.

King, C.J. 1971. Separation Processes. New York, NY: McGraw Hill.

Ladisch, Michael. 1980. Purdue University, Laboratory of Renewable Resources. Personal communication.

Locke, David. 1981. Queens College, Flushing, NY. Personal communication.

Lockeretz, W., et al. 1980. "Maize Yields and Soil Nutrient Levels With and Without Pesticides and Standard Commercial Fertilizers." Agronomy Journal. Vol. 72, January - February, pp 65-72.

Mc Carthy, J.L. 1954. "Alcoholic Fermentation of Sulfite Waste Liquor." In Industrial Fermentations. Underkofler, Leland A., Hickey, Richard J., eds. Vol. 1. New York, NY: Chemical Publishing Co., Inc.

Moriarity, Andrew J. 1981. Biomedical Resources International, Toronto, Canada. Personal communication.

Noyes Development Corporation. 1964. Ethyl Alcohol Production Technique. Pearl River, NY: Noyes Develoment Corporation.

Hedrick, William. 1980. Consulting Engineer, Denver, CO. Personal communication.

Paturau, J.M. 1969. By-products of the Cane Sugar Industry. Amsterdam, The Netherlands: Elsevier Publishing Co.

Paul, J.K. 1979. Ethyl Alcohol Production and Use as a Motor Fuel. Park Ridge, NJ: Noyes Data Corporation.

Pimentel, D., et al. 1979. "Food Production and the Energy Crisis." Science. Vol. 182, p. 444.

Reed, Thomas B., Hedrick, William S. 1980. "Potential Improvements in Alcohol Production and Use." Golden, CO: Solar Energy Research Institute.

Righelato, R.G., Elsworth, R. 1970. "Industrial Applications of Continuous Culture: Pharmaceutical Products and Other Products and Processes." Advances in Applied Microbiology. Vol. 13. New York, NY: Academic Press.

Rosell, R.E., et al. 1974. Insect Control Guide for Corn and Sorghum. Lincoln, NE: University of Nebraska Extension Service. Bulletin 74-1509.

Saeman, J.F. 1954. "Production of Alcohol from Wood Waste." In Industrial Fermentations. Underkofler, Leland A., Hickey, Richard J., eds. Vol. 1. New York, NY: Chemical Publishing Co., Inc.

Schroder, Eugene. 1981. Schroder Alcohol Fuels, Campo, Colorado. Personal communication.

Solar Energy Information Data Bank. 1980. Data base searches. Golden, CO: Solar Energy Research Institute.

Solar Energy Research Institute. 1980. Fuel From Farms: A Guide to Small-Scale Ethanol Production. Golden, CO: SERI.

Stark, W.H. "Alcoholic Fermentation of Grain." 1954. In Industrial Fermentations. Underkofler, Leland A., Hickey, Richard J., eds. Vol. 1. New York, NY: Chemical Publishing Co., Inc.

Texas Department of Agriculture. 1978. "Texas Grain Sorghum." Austin, TX: Texas Department of Agriculture.

Timmermans, J. 1965. "Physico-Chemical Constants of Pure Organic Compounds." Vol. 1. New York, NY: Elsevier Publishing Company, Inc.

Timmermans, J. 1965. "Physico-Chemical Constants of Pure Organic Compounds." Vol. 2. New York, NY: Elsevier Publishing Company, Inc.

U.S. Department of Agriculture. 1980. Small-Scale Fuel Alcohol Production. Washington, D.C.: USDA.

U.S. Department of Agriculture. 1975. Agricultural Statistics 1975. Washington, D.C.: USDA.

U.S. Department of Agriculture. 1978. Agricultural Statistics 1978. Washington, D.C.: USDA.

U.S. Department of Agriculture. 1979. Agricultural Statistics 1979. Washington, D.C.: USDA.

University of Nebraska. 1980. "Ethanol Production and Utilization for Fuel." Lincoln, NE: University of Nebraska, Extension Service.

Winston, S.J. 1980. Energy Incorporated, Idaho Falls, ID. Personal communication.

Winston, S.J. n.d.a. "Summary Report on Gasohol:
Technical Considerations and Opportunities." Idaho
Falls, ID: Energy Incorporated.

Winston, S.J. n.d.b. "Ethanol Fuel: Use, Production
Principles and Economics." Golden, CO: Solar Energy
Research Institute.

Wittmuss, H., et al. 1975. "Energy Requirements for
Conventional Versus Minimum Tillage." Journal of Soil
and Water Conservation. March-April.

4
Butanol

Properties of Butanol

The chemical formula for 1-butanol--referred to in
this book as butanol and also knwn as n-butanol, n-butyl
alcohol, and butyric alcohol--is $CH_3(CH_2)_2CH_2(OH)$, or
C_4H_9OH. Butanol is colorless and is miscible with
alcohol and ether. It freezes at -89.0°C (-192.2°F),
boils at 117.7°C (243.9°F), has a flash point of 35°C
(95°F), a specific gravity of 0.81, a Motor Octane
Number of 83, an autoignition temperature of 365°C
(689°F), and a wt/gal of 6.76 lb at 20°C (68°F) (Hawley
1981; Noon 1981a).

History of Production and Use

Butanol may be produced as a by-product of
petroleum refining or by fermentation of biomass
feedstocks. Because one purpose of this book is to
evaluate alcohol fuels as an alternative to
petroleum-based fuels, the production of butanol from
crude oil is not discussed.

Butanol as a product of fermentation was first
observed by Pasteur in 1862; acetone as a product of the
same fermentation process was observed by Schardinger in
1905. Research in England started in 1909 on acetone
and butanol as potential chemical feedstocks. While
employed by Strange & Graham, Ltd. of England, Weizmann
of Manchester University and Fernbach of the Pasteur
Institute, isolated bacteria capable of producing amyl
alcohol, butanol, ethanol, and acetone from potato
starch. In 1912, Weizmann left Strange & Graham, to
work on his own. He eventually isolated a microorganism
capable of producing higher yields of alcohol and useful
with a variety of subtrates. This microorganism is now

classified as <u>Clostridium</u> <u>acetobutylicum</u> Weizmann
(McCutchan 1954; Compere 1979).

During World War I, a plant in Terre Haute, IN,
produced acetone by fermentation to be used in producing
cordite, an explosive used in weapons. Butanol was
considered a waste material in this process, and storage
and disposal were problems. The plant was closed after
World War I. Producing butyl acetate--a solvent for
nitrocellulose lacquers used in the automobile
industry--from butanol was initiated in the 1920s. The
Terre Haute plant was reopened, and an additional plant
was built in Peoria, IL; this time butanol was the
primary product, and acetone was a minor by-product. In
1927, 96 fermentors were in operation at the Peoria
plant--with a production capacity of 50,000 gallons
each--with corn as the feedstock. Additional plants
were built in locations such as Philadelphia, PA,
Baltimore, MD, and Puerto Rico. Molasses was the
feedstock. Research in the 1930s led to the development
of various microogranisms that increased alcohol yields
(McCutchan 1954).

Butanol Production by Fermentation

Feedstocks for butanol production by fermentation
include (1) sugar crops, (2) starch crops, (3)
cellulose crops, and (4) agricultural, forest, and
municipal wastes. Feedstocks that have been used and/or
experimented with include: corn, molasses, sugar beets,
wheat, barley, rice, Jerusalem artichokes, whey, oat
hulls, corncobs, bagasse, cassava, sulphite waste
liquor, hydrol (residue from manufacture of crystalline
dextrose from corn starch), and municipal waste. The
most commonly used feedstocks have been corn and
molasses (McCutchan 1954).

There are two traditional industrial fermentation
processes that produce butanol: butanol-isopropyl and
butanol-acetone. They are no longer differentiated and
are called neutral solvents fermentation, which is
similar in some ways to ethanol fermentation.

The butanol-isopropyl fermentation commonly used
the microorganism <u>Clostridium</u> <u>butylicum</u>. In addition to
butanol and isopropyl alcohol, the end products of this
fermentation include carbon dioxide, hydrogen, butyric
and acetic acids, and possible traces of acetone and
formic acid. Use of this process was limited (American
Chemical Society 1952). Butanol-acetone fermentation
was the most widely used fermentation process for
producing butanol and is the basis of neutral solvents
fermentation.

The most commonly used bacteria for neutral
solvents fermentation is <u>C</u>. <u>acetobutylicum</u>--an

anaerobic bacteria that grows optimally at 37°C (98.6°F). When this bacteria is used, the fermentation process may be referred to as the Weizmann process, in honor of the developer (McCutchan 1954; Noon 1981a).

The substrate obtained by saccharification of corn is a hexose monosaccharide. C. acetobutylicum can be used to ferment this substrate in 40 - 56 hours. In the laboratory, on a weight basis, 34% of the hexose monosaccharide can be converted into butanol. An approximately 90% - 95% conversion rate can be achieved (Noon 1981b).

Methane digestion should not be confused with the anaerobic digestion that takes place during neutral solvents fermentations. The suggestion that a digestion system producing methane can be easily switched to produce butanol has appeared in popular literature, but this is not true (Noon 1981b).

Whatever substrate is used, the mash is covered and fermented at the required temperature, pH, and time.

After fermentation, butanol recovery is achieved by fractional distillation. The following separation processes are being developed and are at the bench or bench-pilot scale (Griffith 1981):
1. use of molecular sieves;
2. adsorption; and
3. the fualex process.

An additional method--salting out--has been used in the laboratory to enhance the layering effect of butanol for ease in recovery. It has no commercial application, in part because of the high salinity of the residue, which causes disposal problems. Salting out is based on butanol's limited solubility in water. The addition of sodium chloride further decreases the solubility of butanol. A layer of 160-proof butanol will form when it reaches a saturation point of about 8%. This layer can be decanted. The residue comprises water, acetone, butanol, ethanol, salt, and wet stillage (Noon 1981a).

Distillation is a separation process of two or more liquids in solution that is based on their relative volatilities and takes advantage of their different boiling temperatures. Fractional distillation draws off vapors from different levels of the distillation column. Heavier products will be recovered in the lower part of the column, and lighter products in the upper part. Although the desired end product may be recovered in concentrations greater than those of additional by-products, it is not necessarily pure. However, pure butanol is not required for fuel use. This lower-grade butanol is called technical grade or power grade.

The acetone, with an azeotropic boiling point of 56.1°C (133°F), is recovered first. The maximum concentration obtainable is 88.5% at atmospheric

conditions by simple distillation. The bottoms
product--remaining after acetone distillation--consists
of butanol, ethanol, and water. It is passed to the
second distillation column. Because water can only
absorb butanol up to a 7.8% concentration by weight, the
remaining butanol forms an upper phase layer consisting
of an approximately 80% butanol-20% water product, which
can be recovered. The yields and their respective
solutions of concentration depend on the specific system
used. The lower phase contains butanol saturated water,
which is returned to the still. The ethanol will be
distributed between the upper and lower phase according
to standard equilibrium balances. The water remaining
after butanol distillation can be recycled in the system
(Noon 1981a).

Molecular sieves, such as zeolites or synthetic
analogs, separate molecules of different substances in
solution by selective adsorption on the basis of size.
In this case, the smaller water molecules pass through
the sieve bed, are adsorbed into active sites in the
zeolite and retained. The butanol, acetone, and ethanol
remain in the liquid, which passes out of the bed.

Adsorption utilizes standard chemistry procedures
for preferential adsorption of the butanol molecules
over the water molecules.

The fualex process (fuel-alcohol extraction
process) utilizes a surfactant, such as a polyoxylkylene
polymer, and a hydrocarbon, such as gasoline or diesel
fuel, which are added to the fermented beer. An
emulsion is formed between the water layer and the
hydrocarbon layer. The emulsion layer contains the
alcohols and may be removed repeatedly or continuously.
The Chemistry Division of the Oak Ridge National
Laboratory has invented this process (Compere 1980).

The products of this fermentation process are
butanol, acetone, and ethanol. The ratio varies and is
a function of the miroorganism and specific process
used. Additional coproducts are hydrogen, carbon
dioxide, and stillage. The value of the stillage is a
factor of the feedstock and the microorganism used for
fermentation. The amount of hydrogen and carbon dioxide
produced varies considerably and is a function of the
process used (Griffith 1981; Noon 1981a).

The concentration of the butanol in the
fermentation beer affects the fermentation rate and,
therefore, the yield. Upon reaching a concentration of
2% - 3% wt/vol, active fermentation is discontinued
(Griffith 1981). The butanol itself has a toxic effect
on the microorganisms needed for fermentation.

FEASIBILITY CONSIDERATIONS

Feedstocks

Butanol produced by fermentation generally requires hexose or pentose monosaccharide substrates. Sources of hexose and pentose monosaccharides include (1) sugar crops, (2) starch crops, (3) cellulose crops, and (4) agricultural, forest, and municipal wastes.

Feedstock feasibility considerations are the same as those for ethanol. As with ethanol, the use of potential food crops for fuel production raises many issues. However, it appears possible to produce both food and fuel in the U.S. and in several other countries also. Policies to protect and promote national food and liquid fuel self-reliance include:

1. evaluation of liquid fuel feedstocks in relation to food production quotas;
2. development of fuel crops with lower land, water, soil additive, and pesticide requirements than food crops;
3. use of currently non-producing arable land;
4. full exploitation of the feed value of the stillage resulting from alcohol fuel production;
5. exploitation, as possible, of waste product feedstocks without depleting crop residue necessary for maintenance of healthy soil and as a protection against erosion; and
6. reevaluation of export crops to maximize their use, e.g., domestic use of wheat for alcohol fuel products and export of the dried high-protein feed by-product.

As with ethanol production, feedstock selection should include the following considerations:

1. production ease and cost of obtaining monomeric sugar;
2. yield of fermentable sugar per acre;
3. yield of residue from the production process per volume;
4. cost, composition, and quality of residue for potential use as a coproduct, such as animal feed; and
5. appropriateness of land-cultivation practices for feedstock cultivation.

An analysis by the Massachusetts Institute of Technology (MIT) School of Chemical Engineering Practice indicates that the selling price for butanol is most dependent on feedstock costs and by-product credits. Therefore, selection and availability of feedstocks are important elements in plant siting and production feasibility (Strobel 1981).

Market Potential

Butanol has been suggested as a potential liquid fuel for vehicles. Recommendations include the use of technical grade butanol as a blend with gasoline, No. 2 diesel fuel, acetone, and vegetable oil fuels. Technical grade butanol comprises approximately 83.3% butanol, acetone, and ethanol, and 16.7% water. The ratio of butanol, acetone, and ethanol is a function of the production process and microorganisms used. One process yields butanol, acetone, and ethanol in a ratio of 7.75:4:1, for example. This particular technical grade butanol has a higher heating value of 2.39 x 10^7J/L and a lower heating value of 2.21 x 10^7J/L. The estimated Pump Octane Number is 94. This technical grade butanol does result in about 32% lower miles/gal than gasoline (Noon 1980). Engine tests and commercial-scale feasibility analyses have not been conducted.

Industrial grade--pure--butanol is used extensively as a chemical feedstock. It is expected that costs to produce technical grade butanol will be considerably lower than costs to produce industrial grade butanol (Noon 1981b).

There are no operating full-scale butanol fermentation plants using biomass feedstocks in the United States as of mid-1981. There are such plants operating in other countries, including Taiwan and Korea (Noon 1981b). Reports indicate that the process is not energy efficient, however (Smith 1981).

An economic evaluation of a neutral solvents fermentation system has been completed by MIT. A comparison was made between neutral solvents fermentation and ethanol fermentation. The economics of a 150 x 10^6 kg/yr neutral solvents plant were compared with the standard Department of Energy ethanol plant cost study done by Katzen and Associates. The evaluation concentrated on performing a simplified separation process design and laying the groundwork for future calculations.

A 150 x 10^6 kg/yr fermentation-product stream containing butanol, acetone, ethanol, and water was used as the feed to the separation scheme. In preparation for future separation calculations, parameters for the modified Universal Quasi-Chemical (UNIQUAC) algorithm for estimating liquid activity coefficients were developed. The economics of Clostridia fermentations were evaluated. Since the fermentation product mixture exhibits highly non-ideal behavior, including azeotropes and immiscibility, complex methods of separating the product solvents must be included. Azeotropes form between water and butanol; ethanol and 2-propanol; and ethanol and water. A miscibility gap exists between butanol and water, which in large part is the basis for the separation process. This study provides a new separation process design, including a preliminary cost analysis of a neutral-solvents plant. The plant design and criteria included operating 335 days/yr and production of, but not separation of, the ethanol-water azeotrope. The fermentation product studied was a mixture of 95% water and 5% solvents, composed of 67% butanol, 25% acetone, and 8% ethanol. Liquid activity coefficients calculated by the modified UNIQUAC equation predicted the butanol-water miscibility gap using adjustable binary parameters regressed from experimental data. The UNIQUAC equation, with these parameters, can be used to predict liquid phase non-ideal behavior in future separation calculations (Strobel 1981).

The immiscibility region at the azeotrope composition is used to break the butanol-water azeotrope by decanting the mixture into an aqueous-rich phase and butanol-rich phase. The product specifications included 95% recovery of 99.5% pure butanol and acetone, in addition to an ethanol-water azeotrope (85 wt/% EtOH). The four columns were designed using the McCabe-Thiele analysis and standard sizing procedures. Uninstalled equipment costs ($1.196 x 10^6) for a 150 x 10^6 kg/yr neutral solvents plant compared favorably with a previous neutral solvents separation design ($1.482 x 10^6) and are substantially less than those for an ethanol-separation design ($5.21 x 10^6) (Strobel 1981).

All costs except separation equipment and by-product credits were assumed equal for both plants. The selling price required to produce a 15% net return on investment was 48¢/kg for neutral solvents based on a corn feedstock. (Strobel 1981).

This price is less than both the ethanol selling price of 63¢/kg (corn feedstock) and the present neutral solvents market price of 74¢/kg. Additionally, the overall thermal efficiency for a neutral solvents plant (91%) is higher than that for an ethanol plant (62%) (Strobel 1981).

Energy Analysis

Considerations for evaluating the energy balance of producing technical grade butanol from fermentation are similar to those for producing fermentation ethanol. They include:
1. energy requirements for feedstock production;
2. energy requirements for fermentation butanol production;
3. system efficiency, particularly heat transfer potential; and
4. liquid fuel energy balance.

The same mistakes may be made in determining the energy balance for butanol as with ethanol. These are (1) attributing unnecessary costs to the production process, (2) attributing coproduct treatment costs to the cost of butanol production, and (3) ignoring the liquid fuel energy balance.

The energy cost for producing dry land corn as determined for ethanol production is 7.5 million Btu/acre. At a conservative production rate for technical grade butanol of 75 bu/acre that amounts to 100,680 Btu/bu (Wittmuss 1975; Pimentel 1979). Using figures derived from the production of dry land corn in Illinois, the MIT study arrived at a production rate of 95,200 Btu/bu (Strobel 1981). One bushel of corn (25.5 kg) will yield 25% technical grade butanol, which is 6.4 kilograms, equal to 7.9 liters or 1.79 gallons (Noon 1981a). Therefore, the energy cost for feedstock production for technical grade butanol can be determined by the formula:

$$Btu/gal = \frac{Btu/bu \text{ to produce corn}}{Gal/bu \text{ of butanol produced from corn.}}$$

which is 56,245.81 Btu/gal, using the ethanol production figures, or 53,184 Btu/gal, using the MIT figures.

Using the fermentation-distillation process--which utilizes C. acetobutylicum to permit hexose monosaccharide obtained through the saccharification of corn to produce technical grade butanol--requires 112,000 joules to produce 30 grams of technical grade butanol. This model requires--on a kilogram output basis--3.73×10^6 joules to produce a fuel with a higher heating value of 2.88×10^7 J/kg, which is a 7.7:1 ratio. This model is based on ideal conditions; an operational system may have a ratio of 5:1 (Noon 1981a).

The process described in the MIT study utilizes corn as a feedstock. The plant thermal efficiency was determined as follows:

$$\left(\frac{H_{combustion} \text{ (alcohols + dry grains + hydrogen)}}{\text{energy in utilities} + H_{vap} \text{ (steam)} + H_{combustion} \text{ (corn)}}\right)(100)$$

Table 4.1 gives the heat content by J/kg and J/yr, and production for the neutral-solvents plant in the MIT study. The thermal efficiency for this plant is 91% (Strobel 1981).

Table 4.1
Heat Content and Production in MIT Study

Material	Heat Content 10^6 J/kg	Production 10^6 kg/yr	Heat Content 10^{14} J/yr
Butanol	30.16	99.64	36.03
Acetone	30.82	36.25	11.17
Ethanol	29.74	12.27	3.65
Hydrogen	143.5	7.05	11.17
Dried grains	13.39	253.0	33.88
Corn	-	-	92.67
Utilities steam			8.785
15 psia use	2.252	197.	4.44
150 psia	2.004	3.11	0.062

Source: Strobel 1981.

Environmental and Safety Considerations

Potential environmental problems arising from butanol production by fermentation include (1) excessive removal of crop residue from the land, and (2) excessive use of stillage on the land. Crop residues improve soil quality and help prevent erosion. A limited amount of stillage applied directly to the land or through an irrigation system can improve soil quality. Too much stillage, however, will result in highly acid soil and an unpleasant odor.

Potential environmental and health problems arising from butanol use as a motor vehicle fuel include (1) emissions, and (2) spills. Further testing is necessary to determine emission rates of regulated and unregulated subtances. Initial studies indicate regulated emissions using technical grade butanol are equal to or lower than gasoline or No. 2 diesel oil. However, aldehyde emissions, which are unregulated but potentially hazardous, are higher (Noon 1981b).

Butanol is classified as a flammable liquid and attendent precautions must be taken. Such precautions include:
1. adequate ventilation;
2. banning open flames within the immediate vicinity;
3. adequate coverage by fire extinguishers or a sprinkling system;
4. adequate and well marked exit routes;
5. elimination of spark sources; and
6. proper equipment maintenance.

The following system codes--applicable to all the alcohol fuels--can provide safety guidance and should be considered when constructing a butanol production plant: flammable liquids code, electrical code, and the life-safety code. Additional codes and standards applicable to butanol plant design and operation are:

1. Standard for Storage of Flammable and Combustible Liquids on Farms and Isolated Construction Projects--National Fire Protection Association (NFPA) 395;

2. Prevention of Dust Explosions in Industrial Plants--NFPA 321;

3. Basic Classification of Flammable and Combustible Liquids--NFPA 77;

4. Static Electricity--NFPA 30;

5. Flammable and Combustible Liquids Code--Occupational Safety and Health Administration (OSHA) 190.94;

6. Occupational Noise Exposure--OSHA Subpart O;

7. Machinery and Machinery Guarding--American National Safety Institute (ANSI) B31.1;

8. Power Piping--ANSI B31.3;

9. Chemical Plant and Petroleum Piping--Underwriters Laboratory 142;

10. Standard for Steel Above-Ground Tanks for Flammable and Combustible Liquids--American Society of Mechanical Engineers (ASME) Boiler and Pressure-Vessel Code, Section IV and VIII;

11. National Electrical Code--NFPA 70-1978, where applicable (U.S. Department of Energy 1980).

The Occupational Safety and Health Administration of the Department of Labor, in General Industry Standards, revised 7 November 1978, Section 1910.106, provides detailed information on safety standards and equipment specifications for production and storage of flammable liquids (U.S. Department of Labor 1978).

REFERENCES

Beesch, Samuel C. 1952. "Acetone-Butanol Fermentation of Sugars." Industrial and Engineering Chemistry. Vol. 44, no. 7, pp. 1677-1682.

Compere, A.L., Griffith, W.L. 1979. "Evaluation of Substrates for Butanol Production." [In Development in Industrial Microbiology, vol. 20.] Oak Ridge, TN: Oak Ridge National Laboratory.

Compere, A. L., Griffith, W. L., Googin, J. M. 1980. Fuel Alcohol Extraction Technology Commercialization Conference. Oak Ridge, TN: Oak Ridge National Laboratory.

Griffith, William. 1981. Oak Ridge National Laboratory, Oak Ridge, TN. Personal communication.

High Plains Journal. 1980. "Engineer Sees Problems When Ethanol Separates From Water." 17 March, p. 12-B. Dodge City, KS: High Plains Publishers, Inc.

McCutchan, W. N., Hickey, R. J. 1954. "The Butanol-Acetone Fermentations." Industrial Fermentations. Underkofler, L.A., Hickey, R., eds. New York, NY: Chemical Publishing Co., Inc.

Earth Energy. 1980. "Engine Performance Using Butanol." Vol. 2, no. 2, November.

Noon, Randall. 1980. "The Potential of Butanol." Gasohol U.S.A. December.

Noon, Randall. 1981a. "A 'Power Grade' N-Butanol/Acetone Recovery System." Proceedings from Symposium II - Nonpetroleum Vehicle Fuels. Chicago, IL: Institute of Gas Technology.

Noon, Randall. 1981b. Kansas Energy Office, Topeka, KS. Personal communication.

Peterson, W. H., Fred, E. B. 1932. "Butyl-acetone Fermentation of Corn Meal." Industrial and Engineering Chemistry. Vol. 24, no. 2, February, pp. 237-242.

Pimentel, D., et al. 1979. "Food Production and the Energy Crisis." Science. Vol. 182, p. 444.

Prescott, Samuel Cate, Dunn, Cecil. 1959. "The Acetone-Butanol Fermentation," Chapter 13. Industrial Microbiology, 3rd edition. New York, NY: McGraw Hill Publishing Co.

Prescott, Samuel Cate, Dunn, Cecil. 1959. "The Butanol-Isopropanol Fermentation," Chapter 14. Industrial Microbiology, 3rd edition. New York, NY: McGraw Hill Publishing Co.

Smith, Richard. 1981. Silver Engineering, Denver, Colorado. Personal communication.

Strobel, M.K., Bader, J.B. 1981. "Economic Evaluation of Neutral-Solvents Fermentation Product Separation." Oak Ridge, TN: Oak Ridge Station, School of Chemical Engineering Practice, Massachusetts Institute of Technology and Oak Ridge National Laboratory.

Wittmuss, H., et al. 1975. "Energy Requirements for Conventional Versus Minimum Tillage." Journal of Soil and Water Conservation. March-April.

5
Uses of Alcohol Fuels

INTRODUCTION

Overview

Alcohols have two general uses: (1) as liquid fuels either straight or in blends for various engines, and (2) as feedstocks for the chemical industry and for growing single-cell protein. The emphasis in this book is on alcohols used as fuels. Specific potential uses include:

1. spark ignition engine fuels;
2. stratified charge engine fuels;
3. diesel engine fuels;
4. aircraft piston engine fuels;
5. turbine engine fuels;
6. utility boiler fuels;
7. fuels for fuel cells;
8. chemical feedstocks; and
9. single-cell protein feedstock.

The first five engines listed--spark ignition, stratified charge, diesel, aircraft piston, and turbine--are various forms of internal combustion (IC) engines. IC engines are based on the principle that as gases are heated, they expand and explode, providing the power to push a piston connected to a crankshaft. The process begins when the fuel is ignited. It burns in the presence of air, which causes the gases to heat rapidly and expand. Rapid expansion quickly creates an explosion in the confined space within the engine. The force of the explosion pushes the piston, which moves the crankshaft. IC engines are differentiated primarily by the methods in which the air and fuel are mixed and ignition occurs.

The fuel used in IC engines may be a gas, a vapor, or a fine liquid spray derived from crude oils, petroleum distillates, or alcohols. Gasoline, a well-known IC engine fuel, is actually a combination of hydrocarbons produced by refining crude oil, with characteristics that vary as a function of the crude oil from which it is derived and of the refining process. Octane (C_8H_{18}) is sometimes called gasoline, but it really is a hydrocarbon that is a component of gasoline. Octane is used as a standard for measuring certain qualities of gasoline, such as antiknock, which is expressed by the octane number.

Alcohols can be burned with oxygen to give off large amounts of heat; therefore, they have a significant potential value as liquid fuels. In general, the higher alcohols--those with the most carbon atoms--have the highest heating value. This is because more carbon-hydrogen bonds can be broken to form carbon dioxide and water, releasing more energy. However, it is more difficult and expensive to produce the higher alcohols. Therefore, the simpler alcohols--methanol, ethanol, and butanol--are generally favored for use as alcohol fuels.

The use of alcohols as a fuel is generally less hazardous than the use of gasoline because alcohols have less ignition danger, lower regulated emissions, and reduced toxicity.

While ethanol, methanol, and other closely related alcohol fuels are often considered to be similar, they should be evaluated separately in terms of designated use, performance, and fuel economy, because they function differently. Controls, methodologies, and objectives of alcohol fuel studies vary considerably. Therefore, each study must be evaluated individually.

History of Alcohol Fuel Use

Methanol was an established cooking and heating fuel in Europe by the mid-19th century, and was used for lighting until about 1880, when it was replaced by kerosene, which is more luminescent (Reed 1974). Use of ethanol as an industrial chemical feedstock increased after 1855, when the first denaturant law was passed in Great Britain. Earlier, high excise taxes based on ethanol's use as an alcoholic beverage affected the economic feasibility of using ethanol for other purposes. The denaturant laws provided for a substance to be added to the ethanol to make it unfit for human consumption; such laws are still in effect. Ethanol was the first fuel used to power vehicles in the 1880s and 1890s. Henry Ford presumed it would be the fuel of choice for his automobiles during their earliest stages of development. However, petroleum quickly supplanted ethanol because of its relative abundance and minimal cost. Prior to World War I, the compression ratios in engines averaged about 4:1, and the low-octane fuels produced by directly distilling crude oil were well-suited as fuel for these comparatively inefficient machines (Pleeth 1949).

During World War I, more efficient engines were needed as demand for gasoline exceeded the supply of partly developed petroleum resources and the capabilities of the fledgling refinery industry. Refineries at that time could only convert about 25% of the crude oil to gasoline, because of limitations in distillation technology. The most obvious step to improve engine efficiency was to increase compression ratios, but this in turn produced preignition of the fuel-air mixture, or engine knock. Ethanol and methanol again enjoyed a brief popularity as a fuel additive because of their known antiknock properties (Pleeth 1949).

World Wars I and II and their attendant petroleum shortages brought increased methanol and ethanol production and use, primarily as vehicle fuels. Methanol-water blends were injected in supercharged piston aircraft engines, for example (Bolt 1964). Sweden used a 25% alcohol-gasoline blend, and Cuba used a 10% alcohol-gasoline blend (Wiebe 1954). In the mid-1930s, a blend of ethanol and gasoline, marketed as "Agrol," was sold in the Midwestern United States. Today, straight methanol--also called neat methanol or M100--is used as a fuel in race car engines because of its high-performance capability; but these engines are designed for high-speed use. Except for these isolated examples, little attention was paid to alcohol fuels until the 1973 oil embargo by the Organization of Petroleum Exporting Countries. Since then, studies, symposia, and experimental programs concerning alcohol fuels have been initiated in the United States and other countries. Brazil, for example, has made a major commitment to convert to alcohol fuels and alcohol fuel-gasoline blends for use in motor vehicles.

Up until the 1920s, when it was discovered that butyl acetate--an important industrial solvent--could be produced from butanol, butanol was considered a waste product in acetone fermentation. Butanol has had little attention as an alternative liquid fuel until the late 1970s.

Distribution

Distributing ethanol- and methanol-gasoline blends for general motor-vehicle fuel use in the United States could be complicated by the ability of these alcohols to absorb water, which causes phase separation and inhibits engine performance. In a conventional gasoline storage tank, the water settles to the bottom, allowing the fuel to be siphoned off above that level. Since the U.S. gasoline distribution system is a "wet system" permitting the inclusion of some water, alcohols might have to be stored and distributed separately, with blending taking place at the pump. This would add to the cost of alcohol blends. However, ethanol has been transported in volumes ranging from 5,000 to 30,000 barrels for hundreds of miles through conventional gasoline pipelines in the United States (<u>Alcohol Week</u> 1981). Adding 10% - 15% benzol to ethanol can reduce phase separation (Bolt 1964). Butanol does not have the same problems with phase separation.

Brazil has encountered few problems with the distribution of E20 (a blend of 20% ethanol and 80% gasoline), because of higher ambient temperatures. Also, Brazil's distribution system consists almost entirely of road vehicles, whereas the U.S. system includes pipeline, marine, rail, and road components (Berger 1979).

SPARK IGNITION ENGINES

<u>Description of Spark Ignition Engines</u>

Spark ignition (SI) engines are a class of IC engine that uses a spark to ignite fuel inside the engine cylinder. The fuel and air are blended constantly in requisite proportions through a carburetor and are introduced into the cylinder prior to combustion. Carburetors are designed to mix fuel and air into an SI engine in constant proportions with more air than fuel. The carburetor meters the air and fuel flow, which is called the air-fuel ratio. Maximum fuel economy occurs with a lean mixture--a decrease in fuel; maximum power output occurs with a rich mixture--an increase in fuel (Lichty 1951). Before the fuel will burn, it must be vaporized. The carburetor initiates vaporization by atomizing the fuel into a mist. Some evaporation of the fuel occurs in the carburetor, and the energy spent on this cools the mixture. A carburetor properly calibrated for the fuel in use and for adequate distribution to each cylinder gives the best engine performance.

SI engines are either two- or four-stroke engines. Most SI engines in cars, trucks, and tractors have a four-stroke engine cycle--called the Otto cycle--that operates in carefully timed steps. The requirements of this cycle impose some limitations on the type of fuel that can be used. The first stroke occurs as the piston travels down the crankshaft, the intake valve is opened, and a fuel-air mixture is drawn into the cylinder. The second stroke, the compression intake, occurs as the piston, forced back up the cylinder by the crank, compresses the mixture. The third stroke, the firing or combustion stroke, occurs as the piston reaches the top of the compression stroke and the spark plug fires, igniting the mixture. The mixture explodes and the piston is driven down the cylinder. At the bottom of the firing stroke, the exhaust valve opens, the crankshaft turns, and the piston is once again forced up the cylinder. The burned gases are blown out through the exhaust valve.

Generally, with SI engines, the higher the compression ratio, the more efficient the engine. Compression ratios in SI engines are usually limited by the octane number of the fuel and the need to limit specific air pollutants resulting from high-pressure, high-temperature combustion.

A combustible mixture explodes--or autoignites-- when it reaches its flash point. If an improper fuel (one which has a flash pont that is too low) is used, premature fuel ignition could occur. Fuels for SI engines are graded according to octane number, the measurement of their tendency to autoignite. High-octane fuels have less tendency to autoignite than low-octane fuels. The temperature that the mixture reaches during the compression stroke depends upon the maximum pressure attained. As a result, high-compression SI engines require high-octane fuels.

Use of Alcohol Fuels in Spark Ignition Engines

Power output in SI engines is affected principally by the heat value per cubic foot of chemically correct air-fuel mixture and the requisite compression ratio that will not cause combustion knock (Lichty 1951). The characteristics of good SI engine fuels are:

1. high energy content;
2. high octane number;
3. high latent heat of vaporization;
4. rapid rate of flame propagation;
5. phase stability throughout the range of local ambient temperatures and conditions;
6. convenience of processing, storing, transporting, and distributing; and
7. low level of harmful emissions (Agricultural Alcohol Blended Fuel Study Conference 1975).

Methanol. Methanol has a Motor Octane Number (MON) of 98; a specific volume (ft^3/lb) of 6.64; a Btu/lb of 10,260; a chemically correct air-fuel ratio by weight of 1:6.46, and by volume of 1:7.14; a Btu/ft^3 of chemically correct air-fuel mixture of 108 (Lichty 1951). The decrease in calorific value, although not totally offset, is minimized by the increase methanol produces in volumetric engine efficiency and reduction of preignition problems because of its high octane number.

Methanol differs from petroleum fuels in that it has a low specific heat value but high-performance combustion properties. It is capable of producing increased power at higher efficiencies in high-compression SI engines than petroleum fuels can produce in conventional SI engines.

Considerations for using methanol as a vehicle fuel in SI engines include:

1. Heat of vaporization. Methanol's specific heat of vaporization is 1.167 kJ/kg at 25°C (77°F)--3.8 times greater than isooctane. Provision of sufficient heat to vaporize methanol in an IC engine has been a problem, but effective recovery of exhaust heat could improve thermal efficiency.

2. Vapor pressure. When methanol is used alone, the low vapor pressure causes problems with cold starts below 10°C (50°F), but this can be remedied by electrically igniting a small amount of methanol or by using a low vapor-pressure compound blended with methanol. In hot climates, any problems would be minor. In methanol-gasoline blends, the opposite problem occurs--vapor lock at high temperatures. Removing volatile hydrocarbons--those with a low molecular weight--from methanol blends can reduce vapor lock. This is not a problem in colder climates.

3. Heat of combustion. Lower values of combustion comparing fuels to water vapor are usually used, but should be adjusted for different initial temperatures. [Liquid methanol's heat of combustion will vary $\pm$ 0.0812 kJ/mol K; gaseous methanol's heat of combustion will vary $\pm$ 0.0439 kJ/mol K (Hagen 1977])].

4. Stoichiometry. The stoichiometric formula for oxidation of metahnol is:
    $$CH_3OH + 1.5O_2 = CO_2 + 2H_2O.$$

5. Corrosion and wear. The presence of water and salts can increase methanol's corrosiveness. Metals particularly susceptible are zinc, lead, magnesium, aluminum, and copper. Some plastics--such as polyamics and methacrylate--and rubbers may swell or soften. Polyethylene and polyacetal do not appear to be affected.

6. Phase separation. Methanol's solubility in water makes phase separation a problem when using methanol-gasoline blends. It would be

necessary to ensure that the distribution system was free of water. However, straight methanol can include some water.

7. Octane ratings. Estimates of methanol's MON range from 87.4 - 94.6; estimates for the Research Octane Number (RON) range from 106 - 114. Higher compression ratios are possible with these high octane ratings.

8. Emissions. Nitrogen oxides (NO_x) and carbon monoxide (CO) emissions are decreased considerably with methanol when operating lean. Aldehyde emissions are increased, and are related to how well the methanol is vaporized and mixed with air (Paul 1978).

For methanol-gasoline blends up to M20, the MON increases. The higher the initial octane rating, the less the MON increases when methanol is added (Henein 1978).

Although water can cause phase separation in gasoline-methanol blends, injecting water or adding water to the fuel can decrease NO_x emissions. Decreasing manifold pressure causes similar sharp reductions in nitrous oxide (N_2O). Unburned methanol emissions can be four times greater than unburned gasoline emissions, but better fuel preparation can cut this by 80% - 90%. Increasing the compression ratio from 9.7:1 to 14:1 reduces aldehyde emissions by one-half. (Compression ratios up to 16:1 are made possible in racing cars by burning straight methanol.) Particulates and sulfur compounds are totally eliminated by burning straight methanol (Hagen 1977).

When methanol and isooctane were fuel injected and the mixtures diluted with nitrogen in the air, methanol showed 4.9% greater energy density as a liquid fuel with no evaporation and 1.4% less energy density when completely vaporized at 25°C (77°F) (Hagen 1977).

Ethanol. Ethanol has a MON of 99; a specific volume (ft^3/lb) of 6.62; a Btu/lb of 13,152; a chemically correct air-fuel ratio by weight of 1:8.99, and by volume of 1:14.28; a Btu/ft^3 of chemically correct air-fuel mixture of 104 (Lichty 1951). Ethanol is less volatile and has a higher flash point [12.7°C (54.9°F)] than gasoline. Ethanol vapors are much lighter than gasoline vapors, hence disperse readily, rather than concentrating in the lowest confined elevation. A higher concentration is required to form an explosive mixture with air (Brame 1972).

Observations at 0°C (32°F) indicate that the reduction in partial pressure of blends up to E20 improves cold starts. This characteristic does not produce any observed concomitant increase in the incidence of vapor lock (Pleeth 1949).

As a fuel, ethanol has several advantages over gasoline. It is (1) inherently safer, e.g., burning ethanol can be extinguished with water, (2) is less toxic, (3) has a higher octane rating, (4) produces fewer hydrocarbon emissions and fewer particulate emissions, and (5) releases no known carcinogens. Its principal drawback is that its calorific value is only about two-thirds of the value of gasoline, although it appears to perform as well as gasoline.

The proportion of oxygen to carbon in ethanol is less than in methanol; therefore, the heat of combustion, which is directly related to calorific value, is correspondingly higher. The high latent heat of vaporization removes extra heat from the fuel-air mixture transferred from the carburetor to the cylinder. This makes the charge denser than that of straight gasoline, which increases the volumetric efficiency of the engine and yields more power. This occurs because ethanol requires almost three times as much energy to vaporize as gasoline, and the heat expended to vaporize the ethanol leaves the intake mixture at a lower temperature than it does with straight gasoline. The cooler mixture has a greater density, which means a heavier fuel charge is drawn into the cylinder than would be drawn from the same volume of gasoline and air. The heavier fuel charge in turn yields increased power (Pleeth 1949).

This has two effects on engine operation. First, the cooler fuel mixture is more dense, so an effectively greater fuel charge can be introduced into each cylinder. In addition, the cooler fuel mixture increases displacement, a determinant of power. Second, the extra energy needed to evaporate the fuel makes it difficult to start a cold engine when using straight ethanol. It may be necessary to provide a more volatile fuel to start the engine and/or a system for heating the intake manifold in climates where the temperature frequently drops below 4.4°C (40°F).

The fuel-air mixture is distributed to each cylinder of a multicylinder engine through a specially designed manifold that keeps the fuel and air thoroughly mixed. Carburetor and manifold system characteristics are designed for specific fuels. If the engine is intended to run on gasoline, the main jet--the part of the carburetor that meters fuel flow to the engine--and the manifold have a specific diameter relative to the quantity of fuel needed. A main jet with a larger diameter would be required to run an engine on straight ethanol--also called neat ethanol or E100. Proportionately more ethanol than gasoline must be delivered to the engine to compensate for ethanol's lower heating value and to decrease the ratio of air to fuel, since ethanol requires less oxygen to burn than gasoline. It may also be necessary to increase the diameter of the manifold to allow proper distribution of this volumetrically larger fuel charge. Proper carburetor calibration curves for the specific engine are essential (Winston 1980).

Generally, no modifications of valve timing are necessary to convert an SI engine to run on ethanol-gasoline blends or straight ethanol, although larger intake valves are sometimes used. With straight ethanol, ignition timing is generally retarded (set so the spark occurs fewer degrees before top dead center, or TDC). Ethanol burns faster, and less time is required to completely combust all the mixture because its rate of flame propagation is more rapid than that of gasoline. At slower engine speeds, the time it takes to complete the combustion stroke is longer than at faster engine speeds. Therefore, although the timing can be retarded at idle, the degree to which the engine can be retarded must be progressively reduced back toward the normal setting as the engine speed increases (Stockel 1969; Winston 1980).

Ethanol and gasoline cannot form stable blends if they contain appreciable amounts of water. Therefore, anhydrous ethanol must be used for blending with gasoline. Blends up to E20 can be used successfully in SI engines without engine modifications. Such blends usually result in increased power and efficiency. A slight decrease in mileage may occur because of the lower volumetric heating value of ethanol. Blends up to E25 improve the RON and MON in a nearly linear manner. There is less improvement in the octane ratings when alcohol is added to premium gasoline (Bolt 1964).

Plaque and dirt in the fuel system of older vehicles may be loosened by using ethanol-gasoline blends, because ethanol is a better solvent than gasoline. The engine stalling or sluggishness resulting from the reduced fuel flow may be easily corrected by replacing the fuel filter.

Although not necessary, engine adjustments specifically for ethanol-gasoline blends will improve performance and driveability (Nakaguchi 1979).

Spark plugs that fire at a colder temperature and reduce the spark plug gap may be necessary.

The performance of conventional SI engines using straight ethanol fuel can generally be improved by modifications to increase the compression ratio. This happens because the ethanol has a higher octane rating than gasoline. There are several methods--including the use of special pistons and planing the heads to reduce the cylinder volume--for making these modifications.

Butanol. Butanol has certain advantages over the simpler alcohol fuels--ethanol and methanol--for use in motor vehicles. The low vapor pressure of butanol makes it less likely to cause vapor lock in SI engines than the other alcohol fuels. Butanol's latent heat--433 kJ/kg--is only slightly higher than that of octane (C_8H_{18}), which has a latent heat of 360 kJ/kg. Cold starting and warm-up problems sometimes associated with use of the simpler alcohol fuels would not occur with butanol. Butanol also has a higher stoichiometric air-fuel ratio than ethanol or methanol--11.2 for butanol, compared with 8.99 for ethanol, and 6.46 for methanol. Therefore, a higher percentage of butanol can be mixed with petroleum fuels as an extender without affecting engine performance.

Because of butanol's insolubility, it will remain in a blend with gasoline even if water is present. (One problem with ethanol and methanol is phase separation, which occurs when water is introduced into the methanol- or ethanol-gasoline blend, resulting in separation of the gasoline and methanol or ethanol.)

Vehicle Tests: General

Test Purpose: To measure vehicle exhaust emissions, fuel economy, driveability, performance, and octane response.

Organization/Contacts: General Motors Corp., Research Laboratory; N.D. Brinkman, N.E. Gallopoulos, M.W. Jackson.

Dates: 1975.

Vehicles Tested: One 1973 and one 1974 General Motors model; V-8 engines; 7.5 L displacement; automatic transmissions.

Fuels Tested: M10, M25 [methanol, 199.9 proof]; E10 [ethanol denatured with 5% methanol]; light hydrocarbons were removed from the gasoline to prevent vapor lock and increased evaporative emissions.

Test Conditions/Methodology: Both cars had been driven 4,000 kilometers on commercial unleaded gasoline before the test was begun.

Results: Methanol and ethanol performed similarly in most cases. It was found that adding alcohol to gasoline without carburetor modifications decreased (1) CO emissions, (2) volume-based fuel economy, (3) driveability, and (4) performance (i.e., quick acceleration). Aldehyde emissions were not measured. Depending upon the carburetors' air-fuel ratios, hydrocarbon and NO_x emissions and RON varied.

At low speeds, fuel economy was almost identical for all fuels, but at speeds of more than 48 kmph, E10 resulted in greater fuel economy than M10. Richer carburetion resulted in less deterioration of performance and driveability.

Setting the carburetion 6% richer for methanol decreased hydrocarbon, CO, and NO_x emissions; a 15% richer carburetion decreased CO emissions, but NO_x emissions were increased significantly and hydrocarbon emissions slightly (Brinkman 1975).

Test Purpose: To test the effects of alcohol fuels on spark ignition engine lubrication and wear.

Organization/Contacts: Southwest Research Institute, Army Fuels Lubricants Research Laboratory; Ed Owens.

Dates: 1976 - continuing.

Vehicles Tested: 2.3 L Pinto.

Fuels Tested: M100, E10, E100, 89% ethanol-11% water blend.

Test Conditions/Methodology: American Society of Testing and Materials (ASTM) Sequence V-D procedure with the following modifications: (1) increased oil sump volume to 3.78 L (3.55 kg), (2) removed intake air temperature and humidity controls, (3) changed the blowby measurement plumbing slightly, (4) monitored continuously the exhaust oxygen and CO in order to make more frequent air-fuel ratio adjustments, and (5) reduced the stage I and II engine oil in temperature by 8°C (15°F).

Results: Initial results indicate that during periods of low-temperature engine operation methanol causes engine wear. When reacted with the lubricant through blowby, methanol and its combustion products increase the rate of wear of the piston rings and cylinder bore significantly during short-term testing. In addition, there is (1) reduced sludge and varnish deposits, (2) severe oil dilution by methanol and water, and (3) some corrosion of copper materials. E100 and E10 did not show increased wear. The addition of 11% deionized water to E100 showed increased iron wear at temperatures below 65°C (149°F) (Owens 1980).

<u>Test Purpose</u>: To test and evaluate practical solutions to problems identified with the use of straight methanol and ethanol.

Organization/Contacts: University of Santa Clara (California), Mechanical Engineering Department; R.K. Pefley, Joseph Nebolon.

Dates: 1977 - continuing.

Vehicles Tested: California 2.3 L Ford Pintos, computer modeling, and laboratory engines and engine components.

Fuels Tested: M100, E100.

Test Conditions/Methodology: The following are the primary tests that have been conducted as of mid-1981:
1. assessed alternative fuel delivery systems—including (1) venturi carbuetor modified for straight alcohol, (2) the Dresserator variable venturi shock wave carburetor, (3) the WHB accoustical pulse fuel metering system, and (4) the Bosch multi-point fuel injection system—for improving engine performance, engine efficiency, and reducing emissions;
2. conducted thermokinetic combustion simulation by computer modeling with emphasis on performance and emissions as they relate to compression ratio, spark advance, water-alcohol mixtures, equivalence ratio, and exhaust gas recirculation; and pollution formation within the combustion chamber and exhaust system;
3. examined solutions to cold start problems with alcohol fuels;
4. examined the effect of straight alcohol fuels on the formation of photochemical smog;
5. assessed the advantages and disadvantages of raising the compression ratio of a Ford Pinto 2.3 L engine; and
6. characterized and compared ethanol and methanol in engine dynamometer studies and road vehicle tests for the U.S. Department of Energy.

The following tests will be conducted under contract for the National Aeronautics and Space Administration:
1. develop a conversion kit for large fleet use (half the vehicles will use straight methanol, half will use straight denatured ethanol);
2. conduct cold chamber tests of an alcohol vapor injection system;
3. conduct computer modeling of aldehyde

formation in the exhaust system and formic
acid formation in the combustion chamber;
4. develop test procedures for alcohol fueled
vehicles including calibration techniques for
FID hydrocarbon analyzers and determination of
hydrocarbon emission species from methanol,
ethanol, and methanol- and ethanol-gasoline
blends; and
5. recommend changes to the Federal Testing
Procedure (FTP) for correcting miles per
gallon and grams per mile tests for alcohol
fueled vehicles.

Results: The following results have been reported from
the completed tests listed above:
1. improved fuel nebulization and distribution
are necessary for development of a totally
satisfactory, cost-effective fuel delivery
system;
2. higher compression ratios can be used to
increase power and efficiency without
increasing NO_x emissions, as much as 30% water
can be added to the methanol to reduce NO_x
emissions without reducing power and thermal
efficiency (however, aldehyde and hydrocarbon
emissions may be increased), and methanol will
thermally ignite more easily than gasoline;
3. electric heaters appear to be the most
satisfactory solution for cold starts in mild
climates, and fuel additives combined with an
electric heater appear most satisfactory in
cold climates;
4. the evidence indicates that alcohol fuel
exhaust may be less photochemically reactive
than gasoline, however, conclusions cannot be
made as of mid-1981;
5. increasing the compression ratio by
turbocharging is preferable to combustion
chamber modification;
6. methanol rated better than ethanol in
performance, efficiency, and emission tests
(Nebolon 1981; Pefley 1981).

Test Purpose: To test alcohol fuels in fleet
conditions.

Organization/Contacts: California Energy Commission,
Synthetic Fuels Office with the cooperation of
Volkswagen of America, Ford Motor Co., Conoco/Douglas
Corp., and the County of Los Angeles; James D.
Kerstetter.

Dates: 1980 - continuing.

Vehicles Tested: There will be three fleet tests.   Test
1 will use twelve 1980 Ford Pintos--eight with
retrofitted engines for straight alcohol fuel use (four
will use ethanol and four will use methanol); and four
will serve as control vehicles using gasoline.   Test 2
will use 25 1981 Volkswagen of America Rabbits factory
produced for use of straight alcohols.   Of these, 12
will use ethanol, 12 will use methanol, and one will
serve as a control vehicle using gasoline.   Test 3 will
use 40 Ford Escorts designed for use of straight
methanol.

Fuels Tested: E100, M100.

Test Conditions/ Methodology: Four of the retrofit
vehicles in Test 1 have had engines modified to the
higher compression ratio of 12:1.   Other modifications
include removal of the terneplate fuel tank coating,
installation of a special fuel filter, chemical coating
of the carburetor, enlarging carburetor jets and fuel
passages, and installing an electric heating element in
the intake manifold under the primary carburetor barrel
(Kerstetter 1981).

Results: Not available as of mid-1981.

Test Purpose: To test the exhaust and evaporative
emissions of alcohol fuels.

Organization/Contacts: Southwest Research Institute;
Bruce B. Bykowski.

Dates: 1978 - 1979.

Vehicles Tested: 1978 Ford Malibu, 1978 Ford Mustang,
1978 Saab, 1979 Ford Marquis.

Fuels Tested: E10, tert-butyl alchol [TBA, $(CH_3)_3COH$],
methyl tert-butyl ether [MTBE, $(CH_3)_3COCH_3$].

Test Conditions/ Methodology: Evaporative enclosure and
dynamometer tests were conducted in accordance with
FTP.   The vehicles were tested with gasolines containing
the alcohol fuel additives and with the base gasoline.
Indolene and unleaded gasoline were used as base fuels.
The following regulated and unregulated emissions were
tested for: (1) evaporative hydrocarbon emission losses,
(2) gaseous emissions of hydrocarbons, (3) CO, (4) NO_x,
(5) aldehydes, (6) individual hydrocarbons, (7) ethanol,
and (8) tert-butyl alcohol.

Results: Little or no effect on hydrocarbon or NO_x

emissions on a net mass level was observed, and CO
emission rates were significantly reduced with all the
alcohol fuel additives. Evaporative emissions showed a
test-to-test variability, due in part to the design and
location of the vehicle's evaporative emission control
system. No increase in the unregulated emissions was
observed, with the exception that in some tests there
was a significant increase in the emissions of the
specific fuel used, e.g., ethanol, TBA, or MTBE.
Vehicles using any of the alcohol fuels tested can pass
the dynomometer tests according to FTP; but, depending
on the specific vehicle's evaporative emission system,
some vehicles may not meet the Hydrocarbon Evaporative
Emission tests according to FTP (Bykowski 1979).

Vehicle Tests: Methanol

Test Purpose: To examine the combustion and emission
characteristics of methanol and methanol-isopropanol
fuels.

Organization/Contacts: U.S. Environmental Protection
Agency, Industrial Environmental Research Laboratory,
Energy Assessment and Control Division, Combustion
Research Branch; G. Blair Martin.

Dates: Continuing.

Engines Tested: A test stationary combustion engine
with a refractory lined cylindrical chamber 0.267 meters
in diameter and 1.22 meters in length. The burner was a
standard movable block swirl burner.

Fuels Tested: M100, isopropanol, and a blend of 50%
methanol-50% isopropanol.

Test Conditions/Methodology: The nominal heat input was
maintained at 88,000 watts and the combustion air was
115% of theoretical. Liquid fuels were atomized through
high pressure nozzles, and propane was introduced
through a radial hole injector.

Results: The following conclusions were drawn:
1. Alcohol fuels produce lower emissions of
   nitrogen oxides than distillate oil or
   propane.
2. The NO emissions of alcohol fuels increase as
   the percentage of higher alcohols increases.
3. The low NO emissions for alcohol fuels appear
   to function from the presence and level of
   oxygen in the fuel molecule, which can be
   viewed as a diluent carried in the fuel. The

operative mechanism appears to be related to thermal effects of the fuel, latent heat of vaporization, and/or decreased flame temperature; however, chemical effects cannot be totally ruled out.

4. In the hot wall experimental system, NO levels similar to those for methanol, e.g., 65 ppm, could be attained for distillate oil and propane with flue gas recirculation and could be approached at 50% water for distillate oil emulsions.

5. Based on the calculations, the use of methanol and a water-oil emulsion appears to impose a 6% - 7% increase in stack heat loss compared to distillate oil with flue gas recirculation; however, it is anticipated that there will be some compensating factors that can be designed into practical systems to minimize the losses.

6. The emissions of CO, UHC, and particulate from alcohol fuels were generally the same as, or less than, those for the conventional clean fuels tested (propane and distillate oil).

7. From a technical standpoint, methanol appears to be a satisfactory fuel for stationary combustion systems (Martin n.d.).

Test Purpose: To provide data on fuel economy and emissions using methanol-gasoline blends.

Organization/Contacts: Exxon Research and Development Co.; Eric E. Wigg, Robert S. Lunt.

Dates: 1974.

Vehicles Tested: Thirteen cars ranging from 1967 to 1974 models with various carburetion characteristics.

Fuels Tested: M15, and M15 with both butane and pentane purged from the gasoline component to give the blend the same Reid Vapor Pressure rating as regular unleaded gasoline.

Test Conditions/Methodology: Fuel economy and emissions tests, the 1975 FTP vapor lock tests, and an adaptation of the CRC track test procedure.

Results: Vapor lock increased significantly, but the purged M15 caused less energy loss. Excessively lean carburetion caused stretchiness in performance. Phase separation was the most difficult problem. Higher molecular-weight alcohols replacing 15% of the methanol fraction of the blend diminished but did not eliminate the problem. Fuel economy benefits accrued only for older cars with richer carburetion. For example, there was an 8% increase for 1967 cars. CO and hydrocarbon emissions were reduced only for rich carburetion, but NO_x emissions increased. Formaldehyde emissions increased 25% - 50% (Wigg 1974).

Test Purpose: To test performance, fuel economy, and emissions of methanol-gasoline blends.

Organization/Contacts: Lincoln Laboratory and Massachusetts Institute of Technology; T.B. Reed, R.M. Lerner, E. Hinkley, R.E. Fahey.

Dates: 1974.

Vehicles Tested: 5 unmodified cars--a 1969 Ford, 1969 Toyota, 1971 Ford Torino, 1972 Ford Torino, and a 1974 Ford Pinto.

Fuels Tested: M5 to M50.

Test Conditions/Methodology: Cars drove a standard course (including a 6% grade hill) at a constant speed of 50 mph using various methanol-gasoline blends; other tests included a four-mile circular course, and on the 1969 Toyota, 10,000 miles of commuting and long trips.

Results: No adverse performance effects were noted in blends under M20. Blends from M5 to M15 showed increased fuel economy and performance and lowered CO emissions and exhaust temperatures. Lean misfire was a problem with M50, and there was some hesitation during idling with M30 (Reed 1974).

Test Purpose: To evaluate use of methanol in single-cylinder engines.

Organization/Contacts: Exxon Corp. Research Laboratories; W.J. Most, J.P. Longwell.

Dates: 1975.

Vehicles Tested: CFR single-cylinder variable compression engine; CLR single-cylinder oil test engine.

Fuels Tested: Methanol, methanol-water blends.

Test Conditions/Methodology: The CFR engine was set for a compression ratio of 7.82:1 and a spark advance of 18°

BTDC; the CLR engine was tested at three inlet manifold pressures--29, 25, and 20 in.--at each of three engine speeds--1000, 2000, and 3000 rpm.

Results: Tests using the CFR engine did not utilize the full octane advantage of the methanol-water blends. Nevertheless, energy efficiency improvements were demonstrated. The CLR engine tests demonstrated that methanol had consistently better fuel economy than the isooctane base fuel (Most 1975).

Test Purpose: To test the feasibility of using 15% methanol with gasoline.

Organization: Volkswagen Co., West Germany.

Dates: 1975-1976.

Vehicles Tested: 45 Volkswagen automobiles and small vans.

Fuels Tested: M15; gasoline used had a lowered vapor pressure, (46% aromatics, 99.4 RON)

Test Conditions/Methodology: The vehicles were driven 1.5 million miles in ordinary driving conditions over a 1-1/2 year period.

Results: No vapor lock problems resulted from the lowered vapor pressure of the gasoline. Volumetric fuel economy dropped 3% - 6%, but energy economy increased 2% - 5%. Engine wear rates were the same as those for gasoline, although some inexpensive nonmetallic parts were replaced initially. Some driveability deterioration was reported, but without affecting power. Hydrocarbon emissions decreased 20%, CO emissions decreased 40%, and NO_x emissions increased 20% - 50% (Adelman 1979a).

Test Purpose: To test fuel economy, phase stability, driveability, and emissions of methanol-gasoline blends.

Organization/Contacts: U.S. Department of Energy, Energy Technology Center, Bartlesville, OK; J. Allsup.

Dates: 1975-1977.

Vehicles Tested: Fleet of ten 1974 and 1975 vehicles, in which the 1974s had about 10,000 miles of previous use, the 1975s about 2,500 miles. Two of the 1975 vehicles had oxidation catalysts.

Fuels Tested: M10 blended with low vapor-pressure gasoline, M10 blended with Indolene, and Indolene as a baseline fuel.

Test Conditions/Methodology: Phase stability was tested by periodic determinations of water content.   Emissions tests used both federal and California standards.   Six vehicles did not meet emissions standards when tested with Indolene, which is the EPA certification fuel.

Results: Average fuel economy decreased slightly with methanol.   Octane increased much more when methanol was added to low-octane rather than high-octane fuel. Neither corrosion nor abnormal wear were observed. Driveability was acceptable.
    Emissions tests on one vehicle showed a variability of ± 33% for CO, ± 9% for NO_x, and ± 19% for hydrocarbons.   Therefore, general conclusions are difficult to draw (Allsup 1977).

Test Purpose: To solve problems with cold starts when using methanol fuel.

Organization/Contacts: Texas A & M University, Department of Chemical Engineering; William Harris, R.R. Davison.

Dates: 1977.

Vehicles Tested: Mazda Mizer.

Fuels Tested: M100.

Test Conditions/Methodology: Gasoline was burned for starting and warm-up.   The gasoline fuel system was modified with a fuel cutoff switch so methanol vapor could flow to the carburetor automatically a short time after the engine warmed up.   A float-level sensing switch was installed in the carburetor.   As the engine warmed, heat was conducted through a stainless steel jacket in a methanol vaporizing chamber, which was partially filled from a methanol fuel tank when the ignition switch was turned.   When enough methanol vapor had been formed, the car switched to methanol consumption.

Results: Performance was acceptable with a slight dropoff followed by a surge of power as the engine switched from gasoline to methanol vapor.   Since no liquid methanol entered the carburetor, the fuel was distributed more evenly to the individual cylinders (Lindsely 1977).

Test Purpose: To test the combination of methanol and electric batteries in a lightweight vehicle.

Organization/Contacts: Southern Illinois University, Carbondale, IL; R. Archer.

Dates: 1979.

Vehicles Tested: The basic undercarriage was a modified Volkswagen floor plan with additional battery space.

Fuels Tested: Methanol.

Test Conditions/Methodology: The engine was a 10-horsepower, methanol-fueled, air-cooled IC engine, which drove a 36-volt, 100-ampere-output brushless alternator. The electric charge was provided by twelve 6-volt, 258-amp/hr, deep-cycle batteries similar to those used in golf carts and forklifts. The engine charged the batteries until they were fully charged, at which time the engine shut off automatically.

Results: The vehicle averaged more than 100 mpg. The design reduced the need for replacement batteries, decreased reliance on peak electrical usage from utilities, and extended the vehicle range. (Most electric cars need 4 tons of batteries to achieve the range of a car with 100 pounds of gasoline.) The test car extended the range of an electric vehicle by 75% – 120%, depending upon the amount of time the vehicle was parked. Battery lifetime was extended by avoiding full drain, and the battery bank was able to supplement household electrical input during peak hours. Pollution was reduced to 25% of that from a standard IC engine (Archer 1979).

Test Purpose: To test a methanol-hydrogen fuel system in a four-cylinder engine.

Organization/Contacts: Solar Energy Research Institute; Richard Passamaneck.

Dates: Continuing.

Vehicles Tested: A modified four-cylinder engine.

Fuels Tested: Methanol.

Test Conditions/Methodology: Methanol is dissociated to produce hydrogen and CO. The hydrogen-rich gases are burned in a modified IC engine. The major engine

modifications consist of raising the compression ratio
to 14:1 and increasing air combustion (as much as 100%
excess air beyond the stoichiometrically correct
amount).

Results: The dissociation reaction is highly
endothermic, requiring 22% of the lower heating value of
methanol. This energy comes from the engine exhaust.
Thus, the dissociated fuel--hydrogen and carbon
monoxide--has 22% more energy than the original
methanol. The modifications resulted in 15% increased
engine efficiency for the higher compression ratios and
25% - 100% for the excess air combustion (Solar Energy
Research Institute 1981).

Vehicle Tests: Ethanol

Test Purpose: To test performance capabilities of
Gasohol.

Organization/Contacts: Nebraska Agriculture Products
Industrialization Committee; William Scheller.

Dates: 1974.

Vehicles Tested: Thirty-four vehicles, 1973 to 1975
models.

Fuels Tested: Gasohol (a blend of 10% agriculturally
derived anhydrous ethanol and 90% unleaded gasoline;
E10).

Test Conditions/Methodology: Twelve vehicles were run
on Gasohol for the entire test, six were run on unleaded
gasoline, and sixteen were run half the time on gasoline
and half the time on Gasohol.

Results: The study found no significant problems with
Gasohol. There were no problems with phase separation,
vapor lock, or driveability, even during winter.
    Fuel consumption decreased with the use of
Gasohol. Total emissions were somewhat less with
Gasohol than gasoline, with significantly reduced CO
emissions (Scheller 1975).

Test Purpose: To test the performance and fuel economy
of Gasohol.

Organization: North Dakota State University,
Agricultural Engineering Department; Ken R. Kaufman.

Dates: 1978 - 1979.

Vehicles Tested: Two 1976 Ford Torinos with 5.75 L V-8 engines, automatic transmissions; one car had been driven 61,000 kilometers, the other 77,000 kilometers.

Fuels Tested: Gasohol (10% agriculturally derived anhydrous ethanol and 90% unleaded gasoline; E10).

Test Conditions/Methodology: Test records were made for summer and winter driving conditions in North Dakota. Carburetors were adjusted to factory specifications at the beginning of the test.

Results: A 3% decrease in mileage was reported. Ambient temperature appeared to affect mileage during winter, but did not appear to affect fuel economy. No significant vehicle performance problems were encountered (Kaufman 1978; Kaufman 1981).

Test Purpose: To test the use of E10 in a vehicle fleet situation.

Organization/Contacts: New Jersey Department of the Treasury; W.F. Hills.

Dates: 1979 - 1983.

Vehicles Tested: Plymouth Duster, Plymouth Fury, Dodge Dart, Dodge Aspen, Dodge Van, AMC Concord, Chevrolet Chevette, Ford Fairmont; models ranged from 1974 to 1980.

Fuels Tested: E10 (200-proof ethanol and unleaded gasoline).

Test Conditions/Methodology: The cars logged 667,180 miles the first year. Control vehicles of the same type and year were driven 251,456 miles with unleaded gasoline.

Results: Acceleration was better with E10, but stalling was more frequent. Gas filters showed no material degradation from use with E10. In general, unmodified vehicles performed as well on E10 as they did on gasoline. Fuel economy results were mixed. The Duster, Dart, and Chevette had better mileage with E10, while the other models had better mileage with gasoline (Hills 1980).

Test Purpose: To test ethanol-gasoline blends in a
fleet situation.

Organization: U.S. Department of the Interior, National
Park Service, Curecanti Recreation Area; Glen D.
Alexander.

Dates: 1980.

Vehicles Tested: Conventional automobiles, pickup
trucks, four-wheel drive vehicles, vans, trucks, trail
caterpillar, road-sized snow blower, boats, all of which
used E25; and eight Chevrolet Luv small pickups, which
were converted to straight ethanol use.

Fuels Tested: E25 and E100.

Test Conditions/Methodology: The vehicles were used for
a total of 250,000 miles over a 13-month period in
standard and rugged driving conditions (Curecanti is
located in southwestern Colorado in the Rocky
Mountains).

Results: No significant maintenance problems were
encountered. Emission tests showed a decrease with E25
over gasoline of 10% - 75% in hydrocarbons and 10% - 90%
for CO. No measureable effects on fuel economy were
identified with E25 (Rounds 1981).

STRATIFIED CHARGE ENGINES

Description of Stratified Charge Engines

     Stratified charge engines are a variation of the
spark ignition IC engine. There are two basic kinds of
stratified charge engines. The open chamber type mixes
the fuel with the air (a lean fuel mixture) during
mixture formation to generate a stratified charge (Ford
Proco, Texaco TCP, IFP type, and Halbers type). The
divided chamber type uses a prechamber to facilitate
stratified charge formation (Broderson, Heintz Ram,
Volkswagen PCI, and Honda CVCC). The Soviet Union also
manufactures a divided chamber stratified charge
engine. Several stratified charge (fuel-injection)
engines are manufactured under Hesselman patents. In
these engines (1) air is inducted into the engine, (2)
fuel is injected into the compressed air near the end of
the compression stroke, (3) the mixture is stratified at
part load, (4) an air swirl in the combustion chamber
carries the mixture to a spark plug, and (5) the charge
is ignitied (Lichty 1951).

In the first chamber of the divided chamber engine,
a rich air-fuel mixture is injected and ignited by a
spark, and combustion begins. In the second chamber, a
lean air-fuel mixture is injected, and the mixture from
the first chamber is fed into the second chamber. The
mixtures are sparked, and combustion occurs. The rest
of the engine cycle is standard spark ignition engine
procedure. The purpose of using a stratified charge
system is to accomplish more complete burning of the
fuel than in standard SI engines.

Use of Alcohol Fuels in Stratified Charge Engines

Direct-injected stratified charge (DISC) vehicles,
such as Ford Proco and Texaco TCP, have the potential
for greater tolerance of variations in fuel properties
than conventional SI engines. The Ford Proco has
greatly reduced octane requirements, and the Texaco TCP
has no octane requirements if the relative timing of the
fuel injection and ignition is correct.
Both the open chamber and divided chamber
stratified charge engines have multifuel capabilities,
which include blends of gasoline with alcohol or alcohol
with water, as well as gasolines of modified
volatility.

Vehicle Tests: General

Test Purpose: To test the performance of alcohol fuels
in stratified charge engines.

Organization: California Energy Commission.

Dates: 1979.

Vehicles Tested: Four 1978 Honda CVCCs with stratified
charge engines.

Fuels Tested: E10, M5, M10, and gasoline.

Test Conditions/Methodology: One car used fuel blend
E10; one used M5; one used M10; and the fourth, the
control vehicle, used gasoline.

Results: With E10, minor problems were reported with
cold starting, hesitation upon acceleration, surging,
and stalling while cold. E10 was equal to gasoline
after engine warm-up and superior to gasoline when
running from a hot start. Overall driveability was
acceptable with all the blends (California Energy
Commission 1980).

Vehicle Tests: Methanol

Test Purpose: To test emissions and efficiency of
methanol in a stratified charge engine.

Organization/Contacts: University of Wisconsin at
Madison, (Department of Mechanical Engineering; C.P.
Chiu).

Dates: 1979.

Vehicles Tested: Four-cycle Honda with one cylinder,
6.5:1 compression ratio, and a displacement of 170 cm^3.

Fuels Tested: Methanol.

Test Conditions/Methodology: The engine was operated at
2,500 rpm. The engine head was modified to accept
methanol injection nozzles that formed a stratified
charge by injecting methanol into an air-gasoline
mixture.

Results: Satisfactory engine output was attained even
when the amount of methanol injected reached 50% of the
total fuel used. NO_x emissions decreased but
hydrocarbon emissions increased as the percentage of
methanol was increased (Chiu 1979).

DIESEL ENGINES

Description of Diesel Engines

     Diesel engines, also called compression-ignition
engines, are high-compression, IC engines. They may be
two- or four-stroke engines. The four-stroke cycle is
called the diesel cycle. Diesel fuel is injected into
the cylinders when the piston has compressed the air so
tightly that it is hot enough to ignite the fuel without
a spark. Diesels that do not have pressurized air
intake--either by turbochargers or superchargers--
maintain a constant air-flow rate to the cylinders,
regardless of engine speed. This means that no attempt
is made to operate with a ratio of air to fuel for
combustion. In diesels that have pressurized air
intake, an increased air-flow rate to the cylinders
results in higher engine speeds. This occurs because
the compressors are linked to engine speed.
     If the fuel is inside the cylinder during the
compression stroke, it will ignite prematurely.
Therefore, the fuel is injected into the engine at the
end of the compression stroke by a high-pressure

injector. This means the fuel must be pressurized to a pressure greater than that inside the cylinder during the compression stroke. Otherwise, the fuel cannot be injected into the cylinder. This is accomplished by a high-pressure injection pump driven by the engine.

The high compression ratio in a diesel engine gives it a very high fuel efficiency. Diesel engines are well-adapted to heavy work because the unthrottled air intake to the engine permits high fuel efficiency with most load conditions and engine speeds.

The principal measure of diesel fuel performance is the cetane rating, which is a measurement of its ignition value. A minimum acceptable cetane number is 30.

Use of Alcohol Fuels in Diesel Engines

Because alcohol fuels have low cetane numbers, little research has been done on their use in diesel engines. Ethanol, for example, has a cetane rating of 6, and methanol of 35. Diesel fuel ranges from 45 to 55. Alcohol fuels cannot be used straight in unmodified diesel engines. However, there are possible uses for alcohol fuel-diesel blends, particularly in modified diesel engines.

There are important reasons for considering alcohol fuels as diesel fuel supplements. The majority of American farmers use diesel-powered equipment. Ethanol, and perhaps butanol, are fuels that can be produced on-farm or by regional farm cooperatives or joint ventures. In case of diesel fuel shortages, it would be advantageous if tractors and other farm equipment could utilize alcohol fuels, at least as a diesel fuel extender. Therefore, in spite of the limitations, ways to use alcohol fuels in diesel engines are being developed to solve the problem of on-farm fuel availability.

Some considerations for utilizing alcohol fuels in diesel engines are:

1. more alcohol fuel than diesel fuel is required by mass and volume;

2. fumigation--incorporating the alcohol fuels in the intake air--could take advantage of the excess air that would exist when using alcohol fuels;

3. ignition delays of alcohol fuels--due to vaporization and cooling of the mixture--must be compensated for; and

4. diesel fuels serve as lubricants for diesel engines; alcohol fuels do not have the same lubricating qualities.

Ways to utilize alcohol fuel-diesel blends include (1) fumigation; (2) separate injection systems for each fuel, also called dual injection; (3) mixture of the fuels just prior to injection; (4) addition of cetane boosters; and (5) spark assisted combustion initiation (Schrock 1980).

With fumigation, the alcohol fuel is ignited by the injected diesel fuel. Increases in power and thermal efficiency have been documented with fumigation techniques (Adelman 1979).

Dual injection appears to lower emissions. With a full load, higher efficiency at low speeds and lower efficiency at high speeds have been documented (Adelman 1979).

Mixture of the fuels prior to injection requires surfactants to stabilize the alcohol fuel-diesel emulsion. Such surfactants may contribute emissions or otherwise affect the process; therefore, they must be chosen with care. Thermal efficiency may be increased slightly, but power output may fall slightly with emulsion systems (Adelman 1979).

Methanol. Dual fuel injection appears to be the only practical way methanol might be used as a diesel fuel supplement. Diesel fuel is injected to provide ignition, with methanol providing the rest of the charge. Carbon particulates are significantly reduced, and the intake manifold temperature is lowered with methanol over diesel fuel. Other characteristics appear to be similar.

Ethanol. Since ethanol and diesel fuel will not form a stable blend if any water is present in either, it is not practical to premix blends for direct use. Also, the vapors from ethanol-diesel fuel blends are combustible (Rotz 1980).

To use ethanol in diesel engines, an autoignition source must be incorporated. Several approaches are being considered. None is recommended by the manufacturer, so any warranty could be voided as a result.

Since ethanol has a high octane rating, it will not preignite in most diesels if it is aspirated into the cylinder with the intake air. This increases the amount of fuel charged to the cylinder, and unless the amount of diesel fuel injected is reduced, overfueling can occur. The consequences of overfueling range from damage to the drive train to destruction of the cylinder. The power delivered by the engine increases greatly as a function of the amount of carbureted ethanol introduced. The temptation toward continually increasing the ethanol input without a corresponding decrease in diesel fuel could result in extensive engine damage.

Engine damage from ethanol carburetion is not caused only by putting too much fuel in the cylinder. Ethanol is a single compound; therefore, it burns at a single flash point. Diesel fuel, on the other hand, consists of many compounds--each with different flash points--which burn relatively slowly as they are injected into the cylinder. The ethanol charge continues throughout the compression stroke. It burns explosively once ignition is started. As a result, the pressure in the cylinder, increases much more rapidly than the piston can travel down the cylinder. The rate of pressure increase and the maximum pressure that results cause piston failure, collapsed injectors, and stretched head bolts. Therefore, the maximum amount of fuel that is carbureted to a diesel engine should not exceed 20% of the fuel charge, regardless of the diesel-fuel injection rate (rack setting).

A ratio between the ethanol carburetion and the diesel fuel injected must be established. Since the air intake flow is constant in ambient-pressure diesels, it is difficult to maintain this control. However, the air flow does increase with engine speed (which results from increased diesel flow) with turbocharged and supercharged diesels. This would allow some control of ethanol input in relation to diesel fuel. There is a commercially available system for injecting lower-proof ethanol into the turbocharger section of diesel engines. Maximum ethanol flow is limited to protect the engine, but the rack setting should also be reduced.

Ethanol can be directly injected into the engine through a separate high-pressure injection system, provided that diesel, or other high-cetane fuel, is injected through a different system just prior to initiating combustion. The fueling rate should not exceed maximum design pressure for the engine. The modifications, including separate injectors, high-pressure fuel pump, and head taps, are expensive and difficult to do. Small-displacement diesels may not have sufficient room on the head to accommodate a properly positioned additional injector for ethanol. Ethanol fueling rates up to 80% of the charge have been reported, but result in substantially reduced volumetric fuel economy (Schrock 1979).

Although ethanol and diesel will generally not form a stable enough mixture to permit use of premixed blends, it may be possible to mix ethanol and diesel just prior to the time they enter the high-pressure injection pump. This would permit use of the conventional injection system. With the exception of requiring two separate fuel tanks and a mixer, the modifications to the system would be minor (Schrock 1979). This approach is being tested by the Ontario Research Foundation in Ontario, Canada.

Ethanol can be burned in diesels if a stable blend with a high-cetane fuel can be prepared. Experiments have been conducted using additives such as amyl nitrate and ethyl nitrate to enhance autoignition. Since both are explosive compounds, the dangers associated with their use may exceed their value. Another drawback is that both are relatively expensive, and quantities that would be required may not be economically feasible (Winston 1980).

Vegetable oils, such as sunflower oil and soybean oil, are combustible and have sufficiently high cetane ratings--30 to 35--for use in diesel engines by themselves. However, these vegetable oils require more crops than ethanol to produce an equal volume of fuel. Although vegetable oils are miscible with ethanol, such a blend still does not have a cetane rating high enough for use in diesel engines (Schrock 1979).

Difficulties in using cetane boosters include the high cost of nitrogen compounds and the fact that they increase NO_x emissions. Most nitrogen compounds also form hydrogen cyanide and/or ammonia emissions.

When using ethanol in a diesel engine, initiation of combustion is a major problem. The addition of a properly oriented spark plug that causes the ethanol to burn as it is injected into the engine would improve the value of ethanol as a diesel fuel. Some tests using this approach have produced satisfactory results. In most instances, this hybrid engine requires radical modifications to the head, which can only be done in a machine shop. Because orientation of the ignition source is critical, a properly engineered system is required (Winston 1980).

Butanol. In diesel fuel, temperatures below 4.4°C (40°F) can cause separation with ethanol. This problem does not occur with butanol. Butanol serves as an antifreeze for diesel fuels, which are cold-sensitive (High Plains Journal 1980).

The stoichiometric air-fuel ratio for butanol is 11.2:1; for diesel fuel it is 15:1. Therefore, less air is required for butanol combustion, and the engine runs leaner.

Tests reported by the National Alcohol Fuel Producers Association indicate that butanol can be used as a diesel fuel supplement up to 40%. Blends up to B40 with No. 2 diesel fuel have no effect on fuel ignition (Earth Energy 1980). Other tests indicate as much as 50% of diesel fuel can be displaced by butanol (High Plains Journal 1980). However, knock can be a problem with heavy loads using B50 (Earth Energy 1980).

Vehicle Tests: General

Test Purpose: To test alcohol-diesel fuel emulsions
medium-speed diesel engine.

Organization/Contacts: Southwest Research Institute,
Department of Engine and Vehicle Research; Quentin A.
Baker.

Dates: 1978 - 1981.

Vehicles Tested: Two-cylinder model of Electro-motive
Division (EMD) Model 567B engine. The bore and stroke
are 21.6 x 25.4 cm, engine displacement is 9.3
L/cylinder, and the maximum governed speed is 835 rpm.

Fuels Tested: Methanol-diesel emulsion, ethanol-diesel
emulsion.

Test Conditions/Methodology: Stabilized and
unstabilized emulsions of methanol-diesel and ethanol-
diesel fuels were tested. Tests were conducted after
the engine had been warmed-up and checked on No. 2
diesel fuel, the baseline fuel.

Test Results: Maximum alcohol content of the emulsions
and solutions was limited by engine knocking as a result
of the reduction on the cetane number. Engine power and
fuel efficiency were slightly below baseline diesel fuel
levels in the high and mid-speed ranges, but were
somewhat improved at low speeds with use of the
unstabilized emulsions and with ethanol solutions.
However, thermal efficiency of the stabilized emulsions
fell below baseline levels at virtually all conditions
(Baker 1981).

Test Purpose: To test stabilized alcohol-diesel fuel
blends.

Organization: University of Missouri at Rolla; Richard
T. Johnson.

Dates: 1979 - continuing.

Vehicles Tested: Cetane fuel research engine; direct
injection, single cylinder engine with John Deere
combustion system; and turbo-charged, direct injection,
four-cylinder John Deere engine.

Fuels Tested: A variety of base fuels, surfactants and
cosolvents with methanol (with 2% water) and ethanol
(with 5% water).

Test Conditions/Methodology: Basic ASTM property evaluation tests are being used with minor modifications in areas such as heating value and temperature stability (Johnson 1981).

Results: Not available as of mid-1981.

Vehicle Tests: Methanol

Test Purpose: To test the combustion and emission characteristics of methanol in gasoline and diesel engines.

Organization/Contacts: Wayne State University, School of Engineering, Center for Automotive Resarch; Naeim Henein.

Dates: 1974 - continuing.

Vehicles Tested: Cooperative fuel research (CFR) gasoline engine.

Fuels Tested: M10 to M40.

Test Conditions/Methodology: The data obtained are analyzed and correlated in terms of the fuel properties, the fuel-air mixture properties, and other design and operating parameters. The CFR Gas Engine is used to measure the octane number.

Results: The following are the results of this study:
1. motor octane rating increases as the percentage of methanol increases;
2. energy consumption in Btu/hp/hour decreases and thermal efficiency increases as percentage of methanol increases;
3. effective pressure at the governing equivalence ratio decreases as the percentage of methanol increases; and
4. NO_x decreases, hydrocarbons increase, and CO follows the trend of the equivalence ratio (Henein 1981).

Test Purpose: To test the use of methanol as the primary fuel with diesel fuel pilot ignition in a two-stroke locomotive diesel engine.

Organization/Contacts: Southwest Research Institute, Department of Engine and Vehicle Research; C.D. Wood, J.O. Storment.

Dates: 1979.

Vehicles Tested: Two-cylinder model of Electro-Motive Division (EMD) Model 567B engine. The bore is 21.6 cm, the stroke is 25.4 cm, engine displacement is 9.3 L/cylinder, and the maximum governed speed is 835 rpm, the piston ratio was set at 20:1. One cylinder was converted for methanol operation.

Fuels Tested: M100, with diesel fuel providing pilot ignition.

Test Conditions/Methodology: The data obtained are analyzed and correlated for speed and load, fuel rates for both methanol and disel fuel, operating temperatures and pressures, and Bosch smoke.

Results: The methanol-fueled cylinder produced power outputs and thermal efficiencies very close to normal diesel values at rated engine speeds and loads, with methanol fueling rates as much as 95% by Btu input. However, the highest thermal efficiencies with high percentages of methanol fueling were accompanied by pronounced knocking (Wood 1980).

Test Purpose: To test the operation of a dual-fuel, two-stroke cycle, medium speed diesel engine with methanol, high-aromatic naptha, and solvent-refined coal.

Organization/Contact: Southwest Research Institute, Department of Engine and Vehicle Research; J.O. Storment, Quentin Baker.

Dates: 1980.

Vehicles Tested: Two-cylinder model of Electro-Motive Division (EMD) Model 567B engine. The bore is 21.6 cm, the stroke is 25.4 cm, engine displacement is 9.3L/ cylinder, and the maximum governed speed is 835 rpm.

Fuels Tested: Methanol, high-aromatic naptha, and solvent-refined coal.

Test Conditions/Methodology: The engine was set up in a dual-fuel configuration. An engine test cycle was used to simulate notched throttle operation of line-haul locomotive engines.

Results: The use of methanol in a diesel engine presents some special difficulties. Aside from its low cetane number, the poor lubricating properties of methanol and its low volumetric heat content present injection equipment problems which must be solved. Methanol heat inputs in excess of 80% were tested. Thermal efficiencies were approximately equal to baseline diesel fuel levels. Knock-free engine operation was achieved in some cases.

Poor injection characteristics of the pilot injection system are believed to be the cause of performance shortcomings in many cases. Pilot spray impingement on the cylinder head and injection from a poor location in the combustion chamber are the two primary problems (Storment 1981).

Vehicle Tests: Ethanol

Test Purpose: To test the use of an ethanol fumigation system in a turbocharged diesel engine.

Organization/Contacts: University of Wisconsin at Madison, Mechanical Engineering Department; J. Chen, D. Gussert, X. Gao, C. Gupta, and D. Foster.

Dates: 1980.

Vehicles Tested. Four-cylinder turbocharged diesel engine.

Fuels Tested: 160-proof ethanol and 200-proof ethanol.

Test Conditions/Methodology: Ethanol is fumigated by diesel injection into the intake manifold via a low-pressure atomizer.

Results: The fumigation of ethanol resulted in a slight improvement in thermal efficiency at high loads and a small reduction at light loads. Ignition delay and rate of pressure rise increased significantly. A change in ethanol proof from 160 to 200 did not significantly alter engine performance (Chen 1981).

GAS TURBINE ENGINES

Description of Gas Turbine Engines

A turbine does not produce power directly; it utilizes the reaction and/or impulse force of moving fluids, vapors, or gases to move a generator, which produces usable power. Turbines are commonly driven by water, steam, gas, or wind. In the gas turbine, air is forced through a compressor into a combustion chamber, where it is mixed with the fuel and ignited by a spark. Up to this point, the gas turbine is similar to an SI engine. After ignition, the gases expand as they are heated to high temperatures and shoot through the turbine, spinning the protruding turbine wheels, which provides rotary energy. The exhaust gases provide thrust. Some of the rotary energy from the turbine wheels is used to run the compressor. Often, a series of turbine wheels will be used to take advantage of as much potential energy from the expanding gases as possible. Turbines (1) can use various fuels easily without modifications, (2) have lower maintenance than standard SI engines because they have fewer moving parts, (3) are lightweight, and (4) are compact. They do require high temperatures for efficient operation, however.

Use of Alcohol Fuels in Gas Turbine Engines

Gas turbine engines can be used without modifications for alcohol fuels; however, modifications would probably increase performance. Possible modifications include (1) explosion-proofing the engine, (2) doubling the fuel flow rates, and (3) adding lubrication. Corrosion and deposits from sulphur, other metals, and sodium are particularly detrimental to gas turbine engines. Because ethanol and methanol are so easily miscible in water, which could contain sodium salts or other impurities, it is particularly important to keep them free of impure water. If these precautions are taken, the turbines should last three to four times longer between overhauls (Hagen 1977).

The U.S. Department of Energy (DOE) predicts that alcohol fuels could supply all projected utility peaking gas turbine requirements as early as 1987, almost three times the projected requirements by 1990, and about four times the projected requirements by 2000 (U.S. Department of Energy 1979).

United Technologies, Inc. tested an 18-megawatt gas turbine for Florida Power Corp.'s Bayboro plant. Seventy-four percent fewer NO_x emissions were reported. However, CO emissions increased 30 ppm above No. 2 distillate oil. United Technologies concluded that methanol was an ideal fuel if cost and supply problems could be solved (Hung 1977).

General Electric Co. has tested methanol in a combustion chamber for potential use as a gas turbine engine fuel. These tests show a 6% gain in power output with neat methanol and a 12% gain with 20% water and 80% methanol above No. 2 distillate oil. NO_x emissions were reduced 40% (Hagan 1977).

The Garrett Corp. has developed a 548 kilowatt-735 hp continuous (800 hp stand-by) gas turbine engine that runs on a 90% ethanol-10% water fuel. The engines are single-shaft, constant-speed, self-governing turbine engines and are in use in Brazil. The ethanol is produced from cassava (manioc) feedstocks. The NO_x emissions are 25% less than in a comparable 800 hp diesel engine, and the CO emissions are 50% lower (Fallon 1981).

AIRCRAFT PISTON ENGINES

Description of Aircraft Piston Engines

Aircraft piston engines are a variation of the four-cylinder SI engines. In this engine, the four pistons are attached to the crankshaft at one point in the center. The purpose of this design is to save weight and space. The engine functions in the same way as the other SI engines--ignited gases expand and push a piston connected to a crankshaft. Performance is a function of speed, altitude, and manifold pressure (Lichty 1951).

Use of Alcohol Fuels in Aircraft Piston Engines

Union Flights of Sacramento, CA, is testing a Super Cub single-engine aircraft on pure 125-octane methanol. According to reports, the plane gives excellent performance, has a cleaner engine, and emits less exhaust that it does with 100-octane, unleaded aviation gas. In addition, the methanol is cheaper--it cost 80¢/gal and the aviation gas cost $2/gal in early 1981. The engine has been modified to run on methanol and has a constant-speed propeller. Some oil is added for lubrication. The engine burns 10% more methanol that traditional aviation gas (Jain 1981).

UTILITY BOILERS

Description of Utility Boilers

A power plant has four primary components (1) a combustion chamber in which fuel is burned and heat created; (2) a means, such as steam or heated gas, to use the heat to exert a force; (3) a machine called a "prime mover," which converts the pressure of heat energy to mechanical energy; and (4) a storage and distribution system for the energy created.

The combustion chamber, often referred to as the boiler, may be a boiler and furnace or an IC engine.

The IC engine could be any type, such as a turbine or SI engine.

Use of Alcohol Fuels In Utility Boilers

Petroleum distillate fuels are commonly used as boiler fuels. Alcohol fuels can be used as substitutes but certain fuel requirements and characteristics must be examined. These include:

1. Stoichiometry. The energy per unit volume of the stoichiometric mixtures of biomass fuels and currently used heating fuels range from 6.0×10^{-4} for low Btu gas ($CO + 2H_2$) to 8.3×10^{-4} for heating oils. Ethanol is 7.8×10^{-4} and methanol is 7.3×10^{-4}. Therefore, air-flow rates would not differ significantly.

2. Fuel flow rates. The lower volumetric energy of alcohol fuels would require larger pumps and nozzles for required larger flow rates.

3. Vapor pressure and heat of vaporization. In continuous combustion systems, such as those found in power plants, the low vapor pressure and high heat of vaporization of alcohol fuels does not cause ignition or flammability problems. The alcohols may be preheated with waste heat.

4. Convective versus radiative heat. The low luminosity of the alcohol fuels creates more radiative and less convective heat than the currently used boiler fuels. However, alcohol fuels burn cleaner and do not produce soot which lessens heat transfer. This may compensate for the loss of convective heat. Boilers may also be redesigned to take fuller advantage of radiative rather than convective heat.

5. Lubrication and wear. The low lubricosity of the alcohol fuels requires the addition of lubricants. Parts must also be tested for compatability with the alcohol fuels. Methanol, for example, has been shown to be corrosive and to cause wear when used with certain metals and plastics.

6. Emissions. Methanol does not emit sulphur compounds because the sulphur must be removed before processing. The alcohol fuels generally emit less CO and fewer particulates than distillate fuels, and NO_x emissions are about equal. Hydrocarbons are usually not a problem in this type of process because complete combustion usually occurs. Aldehydes appear to be less than 1 ppm because of complete combustion conditions.

7. Explosiveness. Flame arrestors may be advisable when using methanol as a fuel in industrial or home furnaces (Adelman 1979).

In tests using methanol as a boiler fuel, NO_x emissions were 25% and 10% less than for natural gas and No. 6 oil respectively. Modifications were limited to using Y-shaped nozzles to aid full methanol combustion and a centrifugal pump to feed the methanol into the boiler. Soot deposits from previous operation were burned off. However, boiler efficiency was 3% less than with natural gas. CO emissions were less than with natural gas or oil. Aldehydes and acids in emissions were insignificant (Hagen 1977).

It has been suggested that before methanol is produced on a scale large enough to fuel a significant percentage of private vehicles, it would be advisable to convert electrical utility boilers and/or gas turbines to burn methanol (Hagen 1977).

FUEL CELLS

Description of Fuel Cells

Fuel cells are a means of storing potential electrical energy as chemical energy. Unlike traditional storage batteries that store energy within the electrode materials, fuel cells act as converters for energy stored outside the cells. They use a fuel and a reactant--usually oxygen from air. The conversion occurs as a chemical reaction as the fuel is fed to one electrode and the oxygen to the other. The resulting reactions affect the charges, and the conversion is maintained as long as the charge is continued, which in turn is dependent upon a steady input of the fuel and the reactant.

The great advantage of fuel cells is that they convert energy isothermally. Therefore, they are not subject to the limited efficiencies of heat engines. In addition, the cells themselves have no moving parts, and the system very few. Therefore, they are clean, quiet, reliable, and require little maintenance.

Fuel cells may be classified according to: (1) type of construction, e.g., whether they use static or flowing electrolytes; (2) type of electrolyte, e.g., acid, alkali, neutral, or molten salt; and (3) operating temperature.

Use of Alcohol Fuels in Fuel Cells

Many companies are studying fuel cells because of the potential high conversion efficiencies. Sizes vary from 20-kilowatt to 25-megawatt capacities, with working efficiency above 35% - 40%. Fuel cells are also being considered for automotive transportation, but the large weight-to-power: weight-for-power ratio must be overcome. Most hydrogen-air fuel cells require expensive catalysts. Methanol is particularly appropriate for use in such fuel cells because it can be catalytically converted to synthesis gas at relatively low temperatures--250° - 350°C (482° - 662°F)--with little carbon or sulphur residue.

In the 1960s, the Exxon Corp. developed an acid electrolyte for methanol-air fuel cells. The highly soluble methanol tended to diffuse toward the cathode and to oxidize directly with the oxidant. Permeable membranes were used to prohibit the diffusion. More recently the Exxon Corp. joined with the Alsthom Co. of France to develop a commercial methanol-air fuel cell for small-scale and dispersed applications (Hagen 1977).

Methanol can be used directly in fuel cells at temperatures of less than 100°C (212°F). It is miscible with polar electrolytes, unlike other hydrocarbon fuels, and can be stored as a liquid.

CHEMICAL FEEDSTOCKS

Chemical feedstocks are the raw materials for the chemical process industry. The chemicals produced may in turn serve as feedstocks for additional products or may be used as is. Chemical feedstocks are one of the major uses of alcohols today.

Methanol is used (1) in the manufacture of formaldehyde and dimethyl terephthalate, (2) in chemical synthesis to produce methyl amines, methyl chloride, and methyl methacrylate, (3) as an antifreeze, (4) as a solvent, and (5) as a dehydrator for natural gas (Hawley 1981).

Ethanol is used (1) as a solvent, (2) to manufacture acetaldehyde, acetic acid, butadiene, dyes, pharmaceuticals, elastomers, detergents, cleaning preparations, surface coatings, cosmetics, (3) as an antifreeze. Ethylene, an important chemical feedstock, is produced by the dehydration of ethanol, although ethylene is produced today primarily as a by-product of petroleum refining. Examples of uses for ethylene include: (1) the production of polyethylene, polypropylene, vinyl chloride, polystyrene, polyester resins, (2) as a refrigerant, and (3) in the welding and cutting of metals (Hawley 1981).

Butanol is used (1) in the preparation of esters, (2) as a solvent, (3) as a plasticizer, (4) in dyeing, (5) in hydraulic fluids, (6) in detergent formulations, (7) as a dehydrating agent, and (8) in butylated

melamine resins.   It can also be used in synthetic
rubber production (Hawley 1981; Noon 1981).
    Today, methanol, ethanol, and butanol are produced
primarily from petroleum-based feedstocks. These
alcohols produced from non-petroleum based feedstocks
provide an alternative means for supplying the chemical
industry with alcohols for feedstocks.

SINGLE-CELL PROTEIN FEEDSTOCK

    Several species of single-cell algae, bacteria, and
yeasts contain a high percentage of protein. These
microorganisms can be used in animal and poultry feed,
as well as in human food. They do not require the land,
water, and nutrients necessary for conventional
agriculture. It is anticipated that conversion
efficiency will be higher than that of conventional
agriculture. The potential implications for world food
production are enormous.
    While methanol is toxic to humans, it is an
excellent food for algae, bacteria, and yeast. Methanol
is being used in Europe on an experimental commercial
basis as a feedstock for growing single-cell protein.
This protein is replacing milk and soybeans in calf
feeds (Paul 1978; Hagen 1977).

REFERENCES

Adelman, H., et al. 1979a. "End Use of Fluids from Biomass as Energy Resources in Both Transportation and Non-Transportation Sectors." Santa Clara, CA: University of Santa Clara, School of Engineering.

Adelman, H. 1979b. "Alcohols in Diesel Engines -- A Review." In Alcohols as Motor Fuels. Warrendale, PA: Society of Automotive Engineers; 1980.

Agriculutural Blended Fuel Study Conference. 1975. Indianapolis, IN: J.V. Longrock.

Alcohol Week. 1981. Vol. 2, no. 9, 2 March.

Allsup, J.R. 1977. "Experimental Results Using Methanol and Methanol/Gasoline Blends as Automotive Engine Fuel." Bartlesville, OK: U.S. Department of Energy.

Archer, R. 1979. "Methanol Electric Hybrid Vehicle: A Comprehensive Approach." Proceedings of the 1979 National Conference on Technology for Energy Conservation, Tucson, AZ, January 23-25, 1979; pp. 381-386. Information Transfer, Inc.

Baker, Quentin A. 1980. "Alternate Fuels for Medium-Speed Diesel Engines." In SAE Technical Paper Series. Warrendale, PA: Society of Automotive Engineers, Inc.

Baker, Quentin A. 1981. "Use of Alcohol-in-diesel Fuel Emulsions and Solutions in a Medium-Speed Diesel Engine." In SAE Technical Paper Series. Warrendale, PA: Society of Automotive Engineers, Inc.

Baker, Quentin A. 1981. "Use of Alcohol-in-Diesel Fuel Emulsions and Solutions in a Medium-Speed Diesel Engine." In SAE Technical Paper Series. Warrendale, PA: Society of Automotive Engineers, Inc.

Bartlesville Energy Technology Center. n.d. Alcohol Fuels in Diesel Farm Engines. Bartlesville, OK: Bartlesville Energy Technology Center.

Berger, J.E. 1979. "Storage and Distribution of Synthetic Fuels." Newman, M. and Grey, J., eds. Utilization of Alternative Fuels for Transportation, Proceedings of the Symposium at the University of Santa Clara, CA; June 13-19, 1978. New York, NY: American Institute of Aeronautics and Astronautics.

Bolt, J.A. 1964. "A Survey of Alcohol as a Motor Fuel." In Alcohols as Motor Fuels. Warrendale, PA: Society of Automotive Engineers; 1980.

Brame, J.S.S., King, J.G. 1972. Solid, Liquid and Gaseous Fuels, 6th edition. London, England: Edward Arnold Publishers.

Brinkman, N.D., Gallopoulos, N.E., Jackson, M.W. 1975. "Exhaust Emissions, Fuel Economy, and Driveability of Vehicles Fueled with Alcohol-Gasoline Blends." Warrendale, PA: Society of Automotive Engineers. Paper no. 750120.

Bykowski, Bruce B. 1979. "Gasohol, TBA, MTBE Effects on Light-Duty Emissions." San Antonio, TX: Southwest Research Institute.

California Energy Commission. 1980. "Methanol/-Ethanol/Gasoline Blend Fuels Demonstration with Stratified Change Engine Vehicles." Sacramento, CA: California Energy Commission.

Chen, J., et al. 1981. "Ethanol Fumigation of a Turbocharged Diesel Engine." In SAE Technical Paper Series. Warrendale, PA: Society of Automotive Engineers, Inc.

Chiu, C.P., Hong, L. H. 1979. "The Effects of Methanol Injection on Emission and Performance in a Carbureted Spark Ignition Engine." Warrendale, PA: Society of Automotive Engineers. Report no. 790954.

Coon, C.W., Jr. 1981. "Multi-Cylinder Diesel Engine Tests with Unstabilized Water-in-Fuel Emulsions." In SAE Technical Paper Series. Warrendale, PA: Society of Automotive Engineers, Inc.

Cox, F.W. 1979. Component Relationships with Two-Phase Gasoline/Methanol/Water Systems. Bartlesville, OK: U.S. Department of Energy.

Craig, Dudley P., Anderson, Herbert J. 1937. Steam Power and Internal Combustion Engines. New York, NY: McGraw-Hill Book Company, Inc.

134

Earth Energy. 1980. "Engine Performance Using Butanol." Vol. 2, no. 2, November.

Elliott, D. 1978. "Conclusions and Recommendations." Utilization of Methanol Based Fuels in Trasnportation. Toronto, Ontario: Ontario Ministry of Transportation and Communications.

Fallon, John. 1981. The Garrett Corporation, Los Angeles, CA. Personal communication.

Gratch, S. 1978. "Outlook for Performance of Alternative Fuels in Automobiles." Newman, M. and Grey, J. (ed.). April 15, 1979. Proceedings of the Symposium-Utilization of Alternative Fuels for Transportation. University of Santa Clara, CA, June 19-23, 1978. New York, NY: American Institute of Aeronautics and Astronautics.

Gregory, D.P. 1972. Fuel Cells. London, England: Mills & Boon, Ltd.

Hagen, D.L. 1977. "Methanol as a Fuel: A Review with Bibliography." In Alcohols as Motor Fuels. Warrendale, PA: Society of Automotive Engineers; 1980.

Hawley, Gessner, ed. 1981. The Condensed Chemical Dictionary, 10th ed. New York, NY: Van Nostrand Reinhold Co.

Heap, M.P. 1975. "Evaluation of NO_x Emission Characteristics of Alcohol Fuels for Use in Stationary Combustion Systems." Research Triangle Park, NC: U.S. Environmental Protection Agency.

Henein, N.A. 1978. "Flame Propagation, Autoignition and Combustion in Alcohol-Petroelum-Air Mixturs and Other Alternative Fuels." Detroit, MI: Wayne State University. Report No. COO-4486-4.

Henein, N.A. 1981. Director, Center for Automotive Research, Wayne State University, Detroit, MI. Personal communication.

High Plains Journal. 1980. "Engineer Sees Problems When Ethanol Separates from Water." Lander, WY; 17 March.

Hills, W.F. 1980. "State of New Jersey's Gasohol Fleet Test: Summary of First Year's Test Results." Trenton, NJ: New Jersey Departent of Treasury.

Hung, W.S.Y. 1977. "The NO$_x$ Emissions Levels of Unconventional Fuels for Gas Turbines." Journal of Engineering for Power. Vol. 99, no. 4, October.

Jain, Bob. 1981. "Small Aircraft Drinks up Methanol." Denver, CO: The Denver Post; 9 April.

Johnson, Richard T. 1981. University of Missouri at Rolla. Personal communication.

Kaufman, K.R., Klosterman, H.J. 1978. "A Highway Test of Gasohol." Farm Research. Vol 37, no. 1. pp. 18-27, July - August.

Kaufman, Kenton R. 1981. "Highway Fuel Economy of Gasohol and Unleaded Gasoline." Fargo, ND: North Dakota State University.

Keller, J.L. 1979. "Methanol and Ethanol Fuels for Modern Cars." Union Oil Company. Presentation for World Federation of Engineering Organizations.

Kerstetter, James D. 1981. "Senate Bill 620 Alcohol Fuels Program." Sacramento, CA: California Energy Commission.

Klein, H. Arthur. 1966. Fuel Cells: An Introduction to Electrochemistry. Philadelphia, PA: J. B. Lippincott Company.

Lichty, L.C. 1951. "Internal-Combustion Engines." In Marks Mechanical Engineers Handbook, 5th edition. New York, NY: McGraw Hill Book Company.

Lindsley, E.F. 1977. "Methanol Conversion for Your Car?" Popular Science. Vol. 211, no. 2, August.

Martin, G. Blair, Heap, M.P. 1975. "Evaluation of NO$_x$ Emission Characteristics of Alcohol Fuels for Use in Stationary Combustion Systems." Presented at American Institute of Chemical Engineers Symposium.

Martin, G. Blair. n.d. "Assessment of Combustion and Emission Characteristics of Methanol and Other Alternate Fuels." Research Triangle Park, NC: U.S. Environmental Protection Agency.

Menrad, H., Lee, W., Bernhardt, W. 1977. "Development of a Pure Methanol Fuel Car." In Alcohols as Motor Fuels. Warrendale, PA: Society of Automotive Engineers; 1980.

136

Most, W.J., Longwell, J. P. 1975. "Single-Cylinder
Engine Evaluation of Methanol -- Improved Energy Economy
and Reduced NO$_x$." In Alcohols as Motor Fuels.
Warrendale, PA: Society of Automotive Engineers; 1980.

Nakaguchi, N.K., Keller, J.L. 1979. "Ethanol Fuel
Modification for Highway Vehicle Use." Washington,
D.C.: U.S. Department of Energy. Report no.
ALO-EY-76-C-04-363-31.

Nebolon, Joseph. 1981. The University of Santa Clara,
Santa Clara, CA. Personal communication.

Noon, Randall. 1981. "A Power Grade N-Butanol/Acetone
Recovery System." In Proceeding from Symposium II -
Non-Petroleum Vehicle Fuels. Chicago, IL: Institute of
Gas Technology.

Owens, E.C., Marbach, H.W., Frame, E.A., Ryan, T.W.
1980. "Effects of Alcohol Fuels on Engine Wear." In
SAE Technical Paper Series. Warrendale, PA: Society of
Automotive Engineers, Inc.

Owens, Edwin C., Marbach, Howard W., Frame, Edwin A.,
Ryan, Thomas W. 1981. "Evaluating the Effects of
Alcohol Fuels on Engine Wear Through Dynamometer Tests."
San Antonio, TX: U.S. Army Fuels and Lubricants
Research Laboratory.

Patterson, D.J., Bolt, J.A., Cole, D.E. 1980.
"Modifications for Use of Methanol or Metrhanol-Gasoline
Blends in Automotive Vehicles." Energy Conservation.
Ann Arbor, MI: U.S. Department of Energy.

Paul, J.K. 1978. Methanol Technology and Application
in Motor Fuels. Park Ridge, NJ: Noyes Data
Corporation.

Pefley, Richard K. 1981. "Characterization and
Research Investigation of Alcohol Fuels in Automobile
Engines." Santa Clara, CA: University of Santa Clara.
Report no. ME-81-1.

Pleeth, S.J.W. 1949. Alcohol, a Fuel for Internal
Combustion Engines. Londan, England: Chapman and Hall.

Reed, T.B., Lerner, R. M., Hinkley, E. D., Fahey, R. E.
1974. "Improved Performance of Internal Combustion
Engines Using 5-30% Methanol in Gasoline." Proceedings
of the Ninth Intersociety Energy Conversion Engineering
Conference; August 26-30, 1974. San Francisco, CA:
American Society of Mechanical Engineers.

Reed, Thomas. 1978. "Alcohol Fuels." Washington, D.C.: Special Hearing of the U.S. Senate Committee on Appropriations; January 31, 1978, pp. 194-205.

Rotz, Alan, et al. 1980. "Utilization of Alcohol Fuels in Spark-Ignition and Diesel Engines." Agriculture Engineering Information Series. East Lansing, MI: Michigan State University. AESI no. 423, file no. 18.8.

Rounds, Michael. 1981. "Test Shows Gasohol Works in Large Fleet." 19 April. Denver, CO: Rocky Mountain News.

Scheller, W.A., Mohr, B.J. 1975. "Gasohol Road Test Program Second Progress Report." Lincoln, NE: University of Nebraska.

Schrock, Mark. 1980. "Using Alcohol in Engines." Presented at the Mid-central Region Meeting of the American Society of Agricultural Engineers, 21 - 22 March 1980; St. Joseph, MO. St. Joseph, MI: American Society of Agricultural Engineers.

Starkman, E.S., Newhall, H. K., Sutton, R. D. 1964. "Comparative Performance of Alcohol and Hydrocarbon Fuels." In Alcohols as Motor Fuels. Warrendale, PA: Society of Automotive Engineers; 1980.

Stockel, Martin W. 1969. Auto Mechanics Fundamentals. South Holland, IL: The Goodheart-Willcox Company, Inc.

Storment, J.O., Baker, Q.A. 1981. "Dual Fueling of a Two-Stroke Locomotive Engine with Alternate Fuels." In SAE Technical Paper Series. Warrendale, PA: Society of Automotive Engineers, Inc.

Strait, J., Boedicker, J.J. 1979. "Diesel Oil and Ethanol Mixtures for Diesel-Powered Farm Tractors." Warrendale, PA: Society of Automotive Engineers.

Tataiah, K., Wood, C.D. 1980. "Performance of Coal Slurry Fuel in a Diesel Engine." In SAE Technical Paper Series. Warrendale, PA: Society of Automotive Engineers, Inc.

Timourian, H., Milanovich, F. 1979. Methanol as a Transportation Fuel: Assessment of Environmental and Health Risks. Livermore, CA: Lawrence Livermore Laboratory.

U.S. Department of Agriculture. 1980. Small-Scale Fuel Alcohol Production. Washington, D.C.: USDA.

U.S. Department of Energy. 1979. "A Comparison of Projected Electric Utility Peaking Gas Turbine Energy Requirements to Potential Alcohol Fuels Availability: 1980-2000." Washington, D.C.: U.S. Department of Energy.

Wiebe, R. 1954. "Dual Carburation with Alcohol-Water Mixtures and Alcohol Blends." Peoria, IL: U.S. Department of Agriculture.

Wigg, E.E., Lunt, R. S. 1974. "Methanol as a Gasoline Extender -- Fuel Economy, Emissions and High Temperature Driveability." In Alcohols as Motor Fuels. Warrendale, PA: Society of Automotive Engineers; 1980.

Winston Steven. 1980. Energy Incorporated, Idaho Falls, ID. Personal communication.

Wood, C.D., Storment, J. O. 1980. "Direct Injected Methanol Fueling of Two-Stroke Locomotive Engine." Warrendale, PA: Society of Automotive Engineers. Report No. 800328.

Wood, C.D., Storment, J.O. 1980. "Direct Injected Methanol Fueling of Two-Stroke Locomotive Engine." In SAE Technical Paper Series. Warrendale, PA: Society of Automotive Engineers, Inc.

Worthington, Robert M., Margules, Morton, and Crouse, William H. 1968. General Power Mechanics. New York, NY: McGraw-Hill Book Company.

6
Current Research and Development

One advantage of incorporating fermentation ethanol--alone and as a blend with gasoline--into national liquid fuel use is that the production technology is known. Ethanol fermentation is a long-established industry. However, determining the most efficient and effective ethanol production methods with fuel as the end product requires additional research and development. Butanol-acetone fermentation is also an established industrial process. However, production of methanol from biomass feedstocks is not commercialized. Exploiting the commercial potential of methanol and butanol also requires additional research.

Sugar and starch crops are the traditional feedstocks for ethanol production. Cellulose is an inexpensive, abundant material that has potential as a feedstock, but it cannot easily be broken down into the sugars needed for fermentation. Therefore, one focus of research is development of a chemically cost-effective means of converting cellulose materials, also called lignocellulose biomass, to fermentable sugars. Other research and development areas are (1) production economics and energy requirements, (2) improved production techniques for sugar and starch feedstocks, (3) methanol production, and (4) butanol production.

The majority of research focuses on commercializing a process for fermentation of cellulose crops to produce ethanol. Cellulose crops consist of cellulose, hemicellulose, and lignin. The cellulose is a polymer of a 6-carbon sugar, primarily glucose, and provides the structure for the plant, much as the skeletal frame provides structure for the human body. The hemicellulose is a polymer of a 5-carbon sugar, primarily xylose, and acts as a sheath around the cellulose, similar to sheaths of muscles in the human body. The lignin is the glue that holds it all

139

together. Lignin cannot be easily enzymatically
degraded. It may be removed by ethanol or butanol
before the production process starts or removed at the
end as a residue. In either case, it may be burned for
its fuel value.

Research on cellulose feedstocks includes (1) an
examination and evaluation of feedstocks, (2)
pretreatment, (3) hydrolysis, (4) fermentation, (5)
recovery of the alcohol, and (6) microbiological and
genetic work with microorganisms for enzymatic
hydrolysis and fermentation.

Feedstock evaluation includes establishing growth
rates, geographical range, potential supply, and cost
determinants.

The goal of pretreatment is to make the polymeric
sugars in cellulose accessible to acid or enzymatic
hydrolysis. Pretreatment may be chemical, mechanical,
or enzymatic. The two primary forms of chemical
pretreatment are using dilute acid to remove the
hemicellulose, and using a solvent, such as ethanol or
butanol, to remove the lignin. The mechanical method
heats the feedstock to a very high temperature and then
rapidly lowers the pressure. The moisture in the
feedstock is vaporized and breaks up the material,
including its crystalline forms (Blanch 1981). IOTECH
Corp. in Ottawa, Canada, holds patents on this process,
and Stake Technology in Oakville, Ontario, Canada, is
also developing aspects of this process.

Specific pretreatment techniques being tested are
(1) autoclave, (2) steam explosion, (3) solvent
delignification, and (4) cellulose solvent. The
autoclave method, utilizing low heat and a 0.5% - 1.0%
sodium hydroxide solution, appears to be effective in
preparing corn stover for the Massachusetts Institute of
Technology enzymatic hydrolysis and fermentation
process, and may be effective in the wet-oxidation
process. Steam explosion appears to be appropriate for
use in the Masonite process and the plug-flow reactor.
The resultant lignin is in active form and could provide
a by-product credit; however, hemicellulose
deterioration may be a problem. Solvent delignification
commonly uses ethanol or butanol as a solvent to
solubilize the lignin fraction of the feedstock. The
resultant butanol-lignin slurry may have potential as a
by-product, e.g., a diesel fuel or a chemical-polymer
feedstock. Cellulose solvent systems are hampered by
the high cost of the solvent and the difficulty of
effectively recycling the solvent. Complex chemical
structures, such as Cadoxen, and concentrated sulphuric
acid are used to selectively dissolve the cellulose
fraction, yielding a highly active product that can
easily be converted to glucose through hydrolysis
(Douglas 1981b).

Low glucose yields, long reaction times, and high costs have hindered commercialization of acid hydrolysis. Research continues on rapid, high-temperature acid hydrolysis utilizing variations of a plug-flow reactor. Kinetic parameters and optimum time and temperature relationships have been defined for reactors (Douglas 1981b).

Research on enzymatic hydrolysis includes: (1) microbiological and genetic work on the microorganisms needed for hydrolysis, and (2) process development. The primary obstacle for cost-effective enzymatic hydrolysis is development of a pretreatment that permits the microorganisms access to the cellulose and hemicellulose.

A mutant of the fungus Trichoderma reesei (also known as T. viride), called C30, has been developed at Rutgers University by Eveleigh and Montenecourt. It is used extensively in research on enzymatic hydrolysis. MIT is working with Clostridium thermocellum, which will produce the enzyme cellulase at high temperatures. In combination with Clostridium thermosaccharolyticum, it can combine the steps of hydrolysis and fermentation, which is referred to as Simultaneous Saccharification and Fermentation (SSF) (Blanch 1981).

One of the questions facing the research community is whether hemicellulose and cellulose should be kept together during the production process or separated and then hydrolyzed and fermented independently. Cellulose is a recalcitrant polymer to hydrolyze and high temperatures, high pressure, strong acid, or certain enzymes are required to break it down into a monomeric sugar.

The xylose monomeric sugar from hemicellulose cannot be fermented by most yeasts. Rather, an anaerobic bacteria is used. This bacteria can also ferment the glucose from the cellulose. Xylose fermentation may require an additional one to two years of research before commercialization.

In addition to addressing the question of whether to ferment hemicellulose and cellulose together, research in fermentation includes identifying the nutrients that assist the fermentation process and determining the optimal conditions for fermentation. The most plausible method for cellulose fermentation appears to be a continuous cell-recycling process using Saccharomyces cerevisae (Blanch 1981).

Other research focuses on Zymomonas, a bacterial microorganism that grows quickly and produces high yields of alcohol. Research on Zymomonas is being conducted in Australia, as well as the United States.

Experiments are also being conducted with thermophilic organisms that can grow at about 65°C

(149°F); yeast grows at about 30° - 35°C (86° - 95°F).
Raising the temperature at which the microorganisms are
grown can contribute to energy savings, since the
boiling point of ethanol is 78.3°C (172.9°F), a
temperature that must be achieved during distillation.

Four primary methods of alcohol recovery are being
researched. The first is an integrated system of flash
fermentation utilizing vacuum distillation, which is
being studied at the University of California at
Berkeley. It has a low energy requirement but high
capital costs, which can be lowered by using a
satisfactory thermophilic microorganism. The second
uses solvents for extracting ethanol from water. The
third uses membranes for separation. This method
involves high costs for pumps and membranes, and
commercialization presents problems. The fourth uses
desiccants, such as zeolites and corn stover.

Because of high investment-capital requirements,
ethanol production from cellulose feedstocks is probably
only going to be appropriate for large-scale plants
producing from approximately 25 million gal/yr - 50
million gal/yr (Blanch 1981).

Research on the energy balance and economics of
ethanol production generally focuses on evaluation and
analysis of established processes in operating
demonstrations or pilot plants.

Research on improved techniques for production of
ethanol from sugar and starch feedstocks is generally
concerned with small-scale plants. This research
includes (1) examining a controlled continuous process,
and (2) attempts to dry the stillage economically. Many
existing small-scale plants, such as the Schroeder plant
in Campo, Colorado, while not included in this list of
current research and development activities, are
nevertheless contributing information on ethanol
production. The Schroeders, for example, consider their
plant a demonstration model and are continually
experimenting with new techniques appropriate for
on-farm, small-scale ethanol production using grain
feedstocks.

Research on methanol production is examining the
use of biomass feedstocks in the gasification process.
Particular emphasis is being placed on the syngas
modification stage.

Research on butanol production is particularly
concerned with separation of the solvents and yields.

ETHANOL PRODUCTION: USE OF CELLULOSE FEEDSTOCKS

Feedstocks

Project: Wood as a Source of Ethanol

Organization: University of Toronto
Department of Chemical Engineering
Toronto, M5S 1A4, Ontario

Telephone: 416/978-3088; 4905

Contact: Morris Wayman

Dates: 1975 - continuing

Synopsis: A three-step process for preparation of
cellulose for alcohol fermentation has been delineated.
This process includes (1) autohydrolysis, (2) lignin
extraction, and (3) enzyme-assisted fermentation.
Celluloses used have been from aspen, eucalyptus,
corncobs, and wheat straw. During the autohydrolysis
extraction process, stronger sodium hydroxide solutions
than normal appear to have the advantage of opening up
the cellulose lattice further and rendering it more
susceptible to enzymatic attack. Research using various
enzymes and yeasts is continuing. The process has been
tested in a run using 3,000 kg (dry weight) of wood
fiber. Additional work indicates that the lignin
obtained from this process may have medicinal value.
When combined with an artificial sugar--lactulose--and
fed to animals, it prevented gallstones and lowered
blood cholesterol levels (Wayman 1977; Wayman 1981).

Project: Examination of Cotton-Gin Residue as an
Ethanol Feedstock

Organization: Texas Tech University
Department of Chemical Engineering
Lubbock, TX 79409

Telephone: 806/742-3553

Contact: L. Davis Clement

Dates: 1980 - continuing

Synopsis: This research focuses on (1) developing the
most effective fermentation sequence using cellulose
feedstocks for bench-scale and pilot plant testing; (2)
developing a process for ethanol production using
cotton-gin residue as the feedstock; and (3) developing
a model environmental impact statement using cotton-gin

144

residue as the feedstock, mild-acid hydrolysis, and
azeotropic distillation (Clement 1981).

Pretreatment

Project: Activation and Chemical Conversion of Wood
Cellulose

Organization: Mississippi State University
P.O. Drawer FP
Starkville, MS 39759

Telephone: 606/325-2116

Contact: Gary McGinnis

Dates: 1981 - continuing

Synopsis: The objectives are to develop the wet-
oxidation process as a pretreatment to activate
cellulose in woody biomass for subsequent hydrolysis and
fermentation, and as a direct process for converting
wood into low-molecular-weight organic acids. The
project is designed to:
1. determine the effects of time, temperature,
and oxygen pressure on the yields of formic,
acetic, and glycolic acids to maximize the
yield of these products;
2. evaluate the catalytic effect of metal ions
(ferric and ferrous salts, copper and nickel
salts) on the rate of formation of formic,
acetic, and glycolic acids from wood;
3. determine the breakdown products from wet
oxidation of lignin at high temperatures;
4. evaluate the economic effects of the two
processes for converting wood into by-products
via the wet-oxidation process;
5. determine the rate of hemicellulose extraction
from solid wood under varying conditions to
maximize hemicellulose yield from wood;
6. evaluate the susceptibility of the wet-
oxidized material to acid and enzymatic
hydrolysis (the acid hydrolysis studies will
be conducted at the MSU laboratory, while
the enzymatic studies will probably be
conducted at another laboratory); and
7. set up a larger (2 gallon) wet-oxidation
reactor to evaluate two- and three-stage
wet-oxidation processes (Solar Energy Research
Institute 1980a).

Project: Enzymatic Transformations of Lignin

Organization: Virginia Polytechnic Institute and State
University,
Departments of Chemistry,
Chemical Engineering, and Forest Products
Blacksburg, VA 24061

Telephone: 703/961-7673; 6122

Contact: Wolfgang G. Glasser
Philip Hall
S.W. Drew

Dates: 1976 - continuing

Synopsis: The objectives of this project are to:
1.   determine the chemical changes in lignin
during the key steps in lignin biodegradation
by microorganisms in laboratory cultures;
2.   optimize the production of lignolytic enzymes
by microorganisms through systematic
variations of fermentation parameters and
lignin substrates; and
3.   refine the computer-aided techniques for
lignin analysis and characterization,
especially with regard to the development of
routine methods for monitoring microbiological
and enzymatic lignin modifications.
Several fermentation parameters affect the rate and
extent of lignin biodegradation, as reflected in
evaluations of lignin-derived CO_2. Among those
investigated were (1) carbohydrate cosubstrate,
(2) temperature, (3) type and level of nitrogen in the
nutrient, and (4) ratio of nitrogen to carbohydrate
source. In addition, significant changes in the
chemical structure of polymeric lignins were identified
at various levels of decay. The hypothesis that
cell-bound enzymes are involved in the biodegradation of
lignin, and that these enzymes produce diffusible
reduced-oxygen species--such as superoxide--which then
mount a rather unspecific attack upon lignin, is upheld
by the experimental observations made during this
research (Glasser 1980; Glasser 1981).

Project: Improved Chemical Utilization of Wood:
Biological Systems for Lignocellulose Conversion

Organization: U.S. Department of Agriculture - Forest
Service
Forest Products Laboratory
P.O. Box 5130
Madison, WI 53705

146

Telephone: 608/264-5719

Contact: Jerome Saeman

Dates: Continuing

Synopsis: This research focuses on the utilization of
lignin-degrading fungi or their enzymes for
delignification of lignocellulose, and for modifying or
degrading industrial-waste lignins and lignins in pulp.
Characterizing cellulose enzymes from anaerobic bacteria
will also be done, since these enzymes appear to be
superior to those from mesophilic aerobes (Saeman
1981).

Project: Pyrolytic Conversion of Lignocellulose
Materials

Organization: University of Montana
                 Department of Chemistry
                 Missoula, MT   59812

Telephone: 406/243-0211; 6212

Contact: Fred Shafizadeh

Dates: 1977 - continuing

Synopsis: Pyrolysis, the controlled thermal treatment
of materials, provides an efficient process for
converting lignocellulose into a wide range of products,
including sugars, chemicals, solvents, charcoal, and
fuel. The overall objective of this project is to
examine promising routes for converting lignocellulose
materials to useful chemicals through pyrolytic means.
Specific objectives are to:
  1.   investigate the thermal depolymerization of
        cellulose in wood and agricultural materials
        into levoglucosan (an anhydride of glucose);
  2.   examine, in an integrated fashion, the
        pyrolytic processes for producing furfural,
        acetic acid, methanol, and carbon as
        by-products from the hemicellulose and xylan
        components of the raw materials;
  3.   establish the application and utility of the
        novel pyrolytic products, especially
        levoglucosan and levoglucosenone; and
  4.   use chemical and physical treatments to
        increase the susceptibility of cellulose-to-
        enzyme hydrolysis by putting an aqueous slurry
        of cellulose or cellulose materials in an
        attritor to achieve wet milling (Stevenson
        1981).

Acid Hydrolysis

Project: Acid Hydrolysis of Concentrated Slurries of
Cellulose Biomass

Organization: Dartmouth College
               Thayer School of Engineering
               Hanover, NH 03775

Telephone: 603/646-2231

Contact: Hans E. Grethlein
          Alvin O. Converse

Dates: October 1977 - continuing

Synopsis: A continuous plug-flow reactor was built for
acid hydrolysis of cellulose materials. A single pass
hydrolysis at 240°C (464°F) with 1% acid gives 50% - 57%
glucose yield for a slurry concentration of 5% - 13.5%
solids. Glucose and xylose yield maps are developed
that indicate the trade off of product, by-product, and
unreacted product. By-product credits may be able to be
taken for furfural. The plug-flow reactor also has been
used for acid pretreatment to cellulose materials prior
to enzymatic hydrolysis. The yield increase by
enzymatic hydrolysis is 21% - 90% for oak, 58% - 100%
for corn stover, and 60% - 93% for newsprint.
Conclusions indicate that for lower quality, higher cost
substrates, such as wood, enzymatic hydrolysis with acid
pretreatment is a competitive process (Converse 1979).

Project: Continuous Twin-Screw Acid Hydrolysis Reactor
for Waste-Cellulose Glucose Pilot Plant

Organization: New York University
               Department of Applied Science
               26 Stuyvesant Street
               New York, NY 10003

Telephone: 212/598-2472

Contact: Barry Rugg

Dates: June 1977 - continuing

Synopsis: Technical problems such as low glucose yields
and long reaction times have prevented large-scale
usage of acid hydrolysis of waste cellulose.
Experiments carried out since 1977 at New York
University have demonstrated that a rapid,
high-temperature acid hydrolysis can produce glucose
yields in the order of 50% - 60% based on the available

cellulose content. It is a high-throughput acid
hydrolysis process based on the application of a Werner
and Pfleidere twin-screw reactor. This process can
utilize many types of cellulose feedstocks, both wet and
dry. The system has been tested utilizing paper pulp at
10% solids and dry hardwood sawdust at 95% solids, with
high conversion yields and good energy efficiency (Rugg
1980a; 1980b; 1980c).

Project: Conversion of Lignocellulose from Wood into
Useful Chemicals

Organization: North Carolina State University
Department of Wood and Paper Science
Raleigh, NC 27650

Telephone: 919/737-3181

Contact: Irving S. Goldstein

Dates: July 1977 - continuing

Synopsis: Utilizing a strong hydrochloric acid,
hydrolysis converts cellulose to glucose, which may then
be fermented to produce ethanol. The hydrolysis of the
hemicellulose will produce primarily mannose and xylose,
and further conversion of these by-products will produce
ethanol and furfural, respectively. The nature and
quantity of the chemicals derived from the hemicellulose
depend upon the origin of the substrate. For example,
the lignin content of the material being hydrolyzed will
vary, depending on whether the substrate is derived from
annual plants or wood. Projected yields from residual
lignin of 35% pure phenol have been suggested. An
integrated wood-chemicals plant has been designed as
part of this project (Goldstein 1980).

Project: Energy and Chemical Production from Wood
Residues

Organization: U.S. Department of Agriculture - Forest
Service
Forest Products Laboratory
P.O. Box 5130
Madison, WI 53705

Telephone: 608/264-5600

Contact: John Zerbe

Dates: 1980 - continuing

Synopsis: This study will examine rapid, high-temperature dilute-acid hydrolysis with low liquid-to-solid ratios to improve the process and obtain higher yields more economically (Zerbe 1980).

Project: Liquid Fuels and Chemical Production from Cellulose Biomass: Hemicellulose and Cellulose Hydrolysis Recovery, and Pentose Utilization in a Biomass Processing Complex

Organization: Auburn University
               Department of Chemical Engineering,
               Chemistry and Microbiology
               Auburn, AL 36830

Telephone: 205/826-4827

Contact: Robert Chambers   Lon Mathias
          Y.Y. Lee   Tim Placek
          Tom McCaskey   Don Vives

Dates: 1977 - continuing

Synopsis: This program has focused on utilization of hemicellulose. The research has been carried out in two major areas: (1) recovery of hemicellulose and cellulose sugars from biomass, and (2) fermentation of hydrolyzates to produce liquid fuel substances. An efficient hemicellulose hydrolysis process using a percolation reactor has been developed. Hemicellulose sugars can be selectively recovered in almost quantitative yields and high concentration. Feedstocks being tested include red oak, corn stover, sweet sorghum stover, and kudzu. The following kinetic model was used to analyze the integral rate data collected:

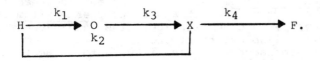

H=hemicellulose
O=soluble oligosaccharides
X=xylose
F=furfural and other decomposition products
k_n=first-order rate constant

    A short-residence-time, high-temperature continuous acid hydrolysis process has been developed.

Research on butanediol production focuses on culture improvement through removal of furfural inhibition. Methods used are heating the medium under vacuum before fermentation, and increasing the rate of cultural inoculation (Solar Energy Research Institute 1980b).

Enzymatic Hydrolysis and Fermentation

Project: GE/CRD Process

Organization: General Electric Co.
              Corporate Research and Development Division
              P.O. Box 8, Building K-1, Room 3B12
              Schenectady, NY   12301

Telephone: 518/385-8640

Contact: R.E. Brooks

Dates: 1977 - continuing

Synopsis: The goal of this research is to demonstrate the technical and economic feasibility of the GE/CRD process for producing ethanol from hardwood and agricultural residues.

In the proposed thermophilic wood-fermentation process, relatively coarse hardwood (and/or softwood) chips are steamed at low pressure in the presence of supplemental amounts of sulphur dioxide (or equivalent reagent) for a brief time, then they are rapidly decompressed. The partially defibrated wood is neutralized with ammonia gas and fed directly into a fermentor, which operates at a temperature of approximately 60°C (140°F). A mixed culture of Clostridium thermocellum and Clostridium thermosaccharolyticum is used to ferment the readily digestible substrate to ethanol. C. thermocellum is used primarily to solubilize cellulose and secondarily to convert cellobiose to ethanol.

C. thermosaccharolyticum is used to ferment the pentose sugars--produced during pretreatment but not utilized by C. thermocellum--to ethanol. Product recovery and cell recycling are accomplished by continuous withdrawal of the broth to a vacuum-distillation chamber and subsequent distillation to produce 95% ethanol (Solar Energy Research Institute 1980b; 1981).

Project: Use of Soluble Cellulose Derivatives as an
Intermediate in Ethanol Production

Organization: State University of New York at Buffalo
Department of Chemical Engineering
Amherst, NY 14260

Telephone: 716/636-2909

Contact: Dane W. Zabriskie

Dates: 1979

Synopsis: The goal of this research is to examine the
feasibility of improving ethanol fermentation utilizing
cellulose feedstocks by chemically modifying the
cellulose to form a water-soluble cellulose derivative,
then homogeneously hydrolyzing this intermediate using
cellulase enzymes. The microorganisms used in this
research are (1) Aspergillus flavus, (2) Pestalotiopsis
westerdijkii, (3) Trichoderma reesei, (4) Myrothecium
verrucaria, and (5) Streptomyces sp. Cellulose sources
are (1) Solka Floc, grade BW-100, (2) sodium
carboxymethyl cellulose, and (3) water-soluble cellulose
acetate. The rate of cellulose hydrolysis and yield of
products are enhanced by using cellulose in a
water-soluble form. The hydrolysis of water-soluble
cellulose acetate using PW filtrate appears to be an
effective system. However, the yield of ethanol is
limited because the anaerobic yeasts cannot assimilate
the acetate groups liberated by the esterase action.
Higher product yields may be obtained with other
microorganisms that can use glucose and acetate as
carbon sources (Zabriskie 1979).

Project: Bioconversion Program

Organization: California Institute of Technology
Jet Propulsion Laboratory
4800 Oak Grove Drive
Pasadena, CA 91103

Contact: Gene Peterson

Dates: 1979 - continuing

Synopsis: The purpose of this research is to develop a
hybrid microorganism from Zymomonas and Pseudomonas to
achieve a one-step process of hydrolysis and
fermentation. The isolation of thermophilic
microorganisms for ethanol production is being
attempted. This would improve the efficiency and reduce

the costs of fermentation ethanol production (Peterson 1981).

Project: Degradation of Cellulose Biomass and Its Subsequent Utilization for the Production of Chemical Feedstocks

Organization: Massachusetts Institute of Technology
Department of Nutrition and Food Science
Cambridge, MA   02139

Telephone: 617/253-2126; 3108

Contact: Daniel I.C. Wang
Charles L. Cooney

Dates: 1977 - continuing

Synopsis: The results of previous work have led to development of a cellulose-to-ethanol process that is being evaluated for potential commercialization. The approach developed at MIT uses a mixed culture of two thermophilic bacteria that grow on both cellulose and hemicellulose, converting them to simple sugars. The organisms then use the 5-carbon sugar and 6-carbon sugar to produce ethanol as the major product. The process has been successfully applied to crop residues such as corn stover. The ability to convert the 5-carbon sugar to ethanol in the same reactor along with the 6-carbon sugar is an economic advantage over most cellulose-to-ethanol processes (Solar Energy Research Institute 1980b).

Project: Enzymatic Degradation of Cellulose Using NOVO Cellulase

Organization: NOVO Laboratories Inc.
59 Danbury Road
Wilton, CT   06897

Telephone: 203/762-2401

Contact: Laurence H. Posorske

Dates: 1977 - continuing

Synopsis: The cellulase employed in this project is obtained from Trichoderma reesei, and is commercially available from NOVO. A kinetic equation describing the action of the enzyme has been developed empirically. Work is in progress to determine the applicability of the equation to a variety of cellulose substrates, including municipal, agricultural, and industrial wastes, and forest products (Lucas 1981).

Project: Enzymatic Degradation of Polymers Produced by Plant and Fungal Cells

Organization: Rutgers University
Cook College
Department of Biochemistry
and Microbiology
New Brunswick, NJ 08903

Telephone: 201/932-9829

Contact: Douglas Eveleigh

Dates: 1977 - 1981

Synopsis: Research at Rutgers University has included an intensive program of enzyme-strain development. The objective is to reduce the manufacturing cost. A series of selective screening methodologies has been designed to isolate mutant strains that can overcome control mechanisms and synthesize high yields of cellulase under conditions that are normally repressive. The purpose of this research is to produce catabolite repression-resistant hypercellulase mutants of Trichoderma reesei, and to clarify the mechanisms controlling the synthesis of cellulase. This project focuses on the microbial conversion of waste agricultural materials into useful by-products. Waste cellulose, such as corn stalks, can be degraded to glucose enzymatically using microbial enzymes, and this product can be fermented to ethanol. The waste agricultural components can be upgraded to high-quality single-cell protein by successive action of microbial enzymes and growth of food yeasts.
Objectives have been to isolate hyperproducing catabolite-repression mutants of the fungus T. reesei. Approximately 100 potential mutants and approximately 800,000 colonies have been examined. Two mutants have been characterized fully.
A mutant series has been obtained that produces high levels of cellulase. One mutant is catabolite repression-resistant to glycerol and glucose for all the enzymes in the cellulase complex. Evaluation of the mutants in fermentors show productivities of 55 - 90 filter paper units/L/hr.
In addition, work is being done with Zymomonas for fermentation. The objective is to develop a strain that will work for both hydrolysis and fermentation. An attempt is being made to transfer both starch- and cellulose-degrading genes to the Zymomonas (Eveleigh 1980).

154

Project: Enzymatic Hydrolysis of Cellulose to
Fermentable Sugars

Organization: U.S. Army Natick R & D Laboratories
               Natick, MA 01760

Telephone: 617/653-1000, Ext. 2091; 2381

Contact: Leo A. Spano
          Mary Mandels

Dates: 1970 - continuing

Synopsis: A prepilot plant with a capacity of
approximately 125 lb/day is in operation. Recent
advances in enzyme production, saccharification, and
pretreatment of materials have brought this particular
process close to economic viability for large-scale
utilization. The enzymes/L have been increased 20-fold
(from 0.85 - 15.0 international filter paper cellulose
units/ml). The enzyme productivity has been raised from
eight international units/L/hr to 100 international
units/L/hr. The hydrolysis rate has been doubled
(production of 10% syrups in 24 hours, instead of 48
hours). Cellulose pretreatment costs have been reduced
by 80% (from 3.5¢/lb to less than 1¢/lb). Based on this
technological progress, a commercial-scale plant capable
of producing 25 million gal/yr of fuel grade (95%)
ethanol would cost $65 million. The factory cost of
ethanol from such a plant is estimated at $1.22/gal.
Projection of process improvements show that by 1983,
the factory cost of 95% ethanol from the same size plant
could be lowered to 75¢/gal (Spano 1979; Mandels
1981).

Project: Integration of Fermentation and Hydrolysis

Organization: Solar Energy Research Institute
               1617 Cole Blvd.
               Golden, CO 80401

Telephone: 303/231-7252

Contact: Karl Grohmann

Dates: 1980 - continuing

Synopsis: The purpose of this research is to develop a
means for integrating the hydrolysis and fermentation
processes for ethanol production. Experiments are being
conducted with the following bacteria: (1) Zymomonas,
(2) Pseudomonas, (3) Escherichia coli, and (4) Bascili
(Grohmann 1981).

Project: Lignocellulose Decomposition by Selected
Streptomyces Strains

Organization: University of Idaho
Department of Bacteriology and
Biochemistry
Idaho Agricultural Experiment Station
Moscow, ID 83843

Telephone: 208/885-6681; 7966

Contact: Don L. Crawford

Dates: Continuing

Synopsis: Several projects concerning the microbiology
of lignocellulose decomposition have been and are being
conducted at the University of Idaho. In one project,
thirty actinomycete cultures were isolated by enrichment
technique on agar media containing newsprint as the
primary carbon and energy source; three Streptomyces
strains were selected for characterization of their
lignocellulose-decomposing abilities. All three
Streptomyces strains were capable of oxidizing
specifically ^{14}CO-labeled lignocelluloses to $^{14}CO_2$.
The Streptomyces attacked primarily the cellulose
(glucan) components, of which between 25% - 40% evolved
as $^{14}CO_2$. Lignin-labeled lignocelluloses were also
attacked, but to a lesser degree, with up to about 3.5%
being oxidized to $^{14}CO_2$. Purified ^{14}C-labeled
milled-wood lignin was also attacked, with recoveries up
to 17.7% as $^{14}CO_2$ (Crawford 1979; Crawford 1980).

Project: Microbial Chemical and Fuel Production from
Fermentation of Cellulose and Starch

Organization: University of Wisconsin at Madison
Agricultural Experimental Station
Department of Bacteriology
116 Agricultural Hall
Madison, WI 53706

Telephone: 608/263-5071

Contact: J.G. Zeikus

Dates: October 1978 - continuing

Synopsis: The purpose of this study is to identify the
optimal organisms and experimental conditions for the

microbial conversion of cellulose and starch to ethanol and acetic acid. New species of thermophilic, anaerobic, and saccharolytic bacteria will be isolated and characterized. Cultural parameters optimal for ethanol and acetate production by Clostridium thermocellum will be determined. The catabolic pathway and its regulation will be studied. The effect of varying cellulose feed rates, source and supply of exogenous growth factors, temperature, and pH--on both production yield and rate of ethanol and acetate production by C. thermocellum--will also be studied.

Bacterial-enrichment cultures will be initiated to isolate new species of thermophilic, anaerobic, and saccharolytic bacteria that metabolize starch, cellulose, and glucose, and that are resistant to high levels of ethanol and acetic acid. Species will be taxonomically identified, and the yield of ethanol and acetic acid from energy sources determined (Zeikus 1981).

Project: Microbial Production of Higher Alcohols and Lipids

Organization: Solar Energy Research Institute
1617 Cole Blvd.
Golden, CO 80401

Telephone: 303/231-1432

Contact: Ruxton Villet

Dates: 1981 - continuing

Synopsis: The project's overall objectives are to:
1.   Develop organisms for very rapid conversion of cellulose, including hemicellulose, and to integrate this conversion with related process developments into an advanced process to convert forage crops--such as sweet sorghum (brown midrib strain)--into alcohol at high yield.
2.   Develop organisms for the efficient conversion of starch-whey mixtures to alcohol fuel and by-product protein feed. Current organisms for converting whey are very slow relative to normal sugar fermentation because whey sugars are about 50% glucose and 50% galactose. The latter sugar is resistant to fermentation. Organisms capable of direct starch-to-alcohol and efficient galactose-to-alcohol fermentation.

are being developed. The objective is to
develop an efficient process that utilizes
large quantities of waste.
3. Screen for microorganisms that can synthesize
higher alcohols (C_3 to C_7 and higher),
determine the parameters influencing higher
alcohol production, and develop a fermentor for
production of a specific product (Solar Energy
Research Institute 1981).

Project: The Microbiology and Physiology of Anaerobic
Fermentation of Cellulose

Organization: University of Georgia
Department of Biochemistry
Center for Biological Resource Recovery
Athens, GA 30602

Telephone: 404/542-1334; 7640; 7424

Contact: Harry D. Peck Jr.
Lars Ljungdahl

Dates: 1979 - continuing

Synopsis: This research focuses on obtaining the
information required to understand, control, and
formulate mixtures of anaerobic bacteria capable of
degrading cellulose to chemical feedstocks, hydrogen,
methane, acetate, and ethanol.
    Anaerobic bacteria are being isolated. These
isolates, plus known laboratory strains, are being used
to partially reconstitute highly active cellulose
fermentations. These mixed cultures will be utilized as
model systems to study the parameters required for the
maximum production of CH_4, H_2, and chemical feedstocks
from cellulose. The physiology of these reconstituted
cultures will be investigated in relation to (1)
cultural conditions, (2) microbial types, (3) inoculum
size, (4) interspecies H_2 transfer, (5) specific
regulatory phenomena, and (6) accumulation of cellobiose
and acetate.
    A major effort will be made to isolate
microorganisms from extreme environments in order to
utilize growth conditions that will enhance the rate of
formation, yield, and recovery of products. When
additional information is available regarding the
microbial types and physiology of reconstituted
cultures, it should be possible to "tailor-make"
microbial populations with regard to substrates,
conditions, and desired products.

The biochemistry, physiology, and bioenergetics of the various types of bacteria included in the fermentation will also be investigated. Emphasis will be placed on (1) the mechanisms and bioenergetics of acetate formation by species of Clostridium and Acetobacterium, and (2) the bioenergetics of sulphite reduction and interspecies H_2 and electron transfer and proton translocation by the H_2 utilizing methanogenic bacteria (Ljungdahl 1981).

Project: Process Development Studies for Bioconversion of Cellulose and Production of Ethanol

Organization: University of California at Berkeley
               Lawrence Berkeley Laboratory
               Berkeley, CA 94720

Telephone: 415/486-5161

Contact: Charles R. Wilke
          Harvey W. Blanch

Dates: September 1976 - continuing

Synopsis: This project has concentrated on hydrolysis of cellulose and hemicellulose to sugars, and their subsequent fermentation to ethanol. A range of cellulose raw materials, principally in the form of agricultural and forest product residues and whole-tree biomass, have been analyzed for chemical composition and assessed for potential yields of sugars and ethanol under various processing conditions. Processing concepts under study include chemical pretreatment of cellulose materials to remove lignin and hemicellulose (pentose sugars), enzymatic and chemical hydrolysis of cellulose to glucose (hexose), and conversion of pentose and hexose to ethanol by fermentation.

Important basic problems include (1) kinetics of enzymatic hydrolysis, (2) physical and chemical nature of the raw materials, (3) production of hydrolyzing (4) enzymes, enzyme absorption as a basis for potential recovery processes, and (5) microbial systems and techniques for fermentation of carbohydrate to ethanol. Major areas of recent progress include (1) development of a highly effective continuous process for production of a cellulase enzyme from the fungus Trichoderma reesei, (2) new ethanol fermentation methods employing dense yeast cultures and operation under vacuum, and (3) operation of major parameters governing the fermentation of xylose to ethanol.

Specific studies include research into (1) the hydrolysis process, (2) production of cellulase enzymes

from high-yielding mutants, (3) enzyme recovery in
cellulose hydrolysis, (4) countercurrent hydrolysis
reactor development, (5) xylose fermentation, (6)
xylanase production, (7) vacuum fermentation and
distillation, (8) low-energy separation process, and (9)
large-scale hollow-fiber reactor development (Blanch
1981).

Project: Thermophilic Fungi-Derived Cellulase for
Solubilization of Cellulose Wastes

Organization: Stanford Research Institute
                Life Sciences Division
                333 Ravenswood Ave.
                Menlo Park, CA   94025

Telephone: 415/326-6200, Ext. 5080

Contact: W.A. Skinner

Dates: 1976 - continuing

Synopsis: Purification of the C_1, C_x, β-glucosidase
enzymes in the extracellular fluid from the culturing of
a thermophilic fungus has been accomplished. Several
separate enzymes from each of the three classes have
been partially resolved, and some have been purified and
studied. Ways to increase the yield of these enzymes
have been explored, and some success has been achieved.
A patent has been issued, and SRI is seeking licenses
and research funds to develop the necessary data to
evaluate the economic and commercial potential for
ethanol production. Comparisons have been made in
temperature, pH, substrate optima and stablility
between the thermophilic fungus, Thielavia terrestris,
and Trichoderma reesei. An additional aspect of this
research is an examination of methods of enzyme
production using recombinant DNA technology (Tuse
n.d.).

Project: Utilization of β-glucosidase in Enzymatic
Hydrolysis of Cellulose

Organization: University of Connecticut
                Department of Chemical Engineering
                Storrs, CT   06268

Telephone: 203/486-4019; 2756

Contact: Herbert Klei
           Donald W. Sundstrom
           Robert Coughlin

Dates: 1980 - 1981

Synopsis: The immobilized enzyme technology already
developed is being applied to cellulose hydrolysis
reactors.   Since the cost of enzyme production
represents a large part of the overall alcohol
production cost, enzyme immobilization offers a way of
reducing the alcohol cost by allowing the enzymes to be
easily recovered.   In addition, enzyme immobilization
allows the enzyme mixture to be monitored in a reaction
system rather than in a mixture supplied by a specific
organism.   Conclusions drawn from this research are:
1.   β-glucosidase from <u>Aspergillus</u> <u>phoenicis</u> was
     adsorbed on controlled pore alumina to yield
     an immobilized enzyme with high activity
     retention and long life under reaction
     conditions;
2.   the immobilized β-glucosidase was
     characterized with cellobiose as the
     substrate, giving an activity maximum near pH
     3.5, an activation energy of about 11 kcal/mol
     and an apparent Michaelis constant of about
     2.7 mM; and
3.   the presence of immobilized β-glucosidase
     during enzymatic hydrolysis of cellulose
     materials significantly increased the
     concentrations of glucose by converting
     cellobiose effectively to glucose (Sundstrom
     1981).

Alcohol Recovery

Project: Development of Ethanol-Selective Membranes

Organization: Southern Research Institute
                2000 Ninth Ave. South
                Birmingham, AL   35205

Telephone: 205/323-6592

Contact: Robert E. Lacey

Dates: 1981 - 1982

Synopsis: Ethanol-selective membranes that combine high
selectivities for ethanol over water with low
resistances to the transfer of the permeating species
are being developed.   The approach involves the
formation of a thin, ethanol-selective layer on the
surface of a low-resistance substrate, either by a
method involving direct chemical treatment of the
low-resistance substrate or by a method involving

161

ultrafiltration of ethanol-selective polymers in and
onto a substrate followed by cross-linking to effect
firm attachment (Solar Energy Research Institute 1981).

Project: Membrane Development for Low-Energy
Separations

Organization: Solar Energy Research Institute
                1617 Cole Blvd.
                Golden, CO 80401

Telephone: 303/231-1000

Contact: Paul Schissel

Dates: 1981 - continuing

Synopsis: Membrane separation technologies offer the
potential for economical and energy efficient separation
of chemical products from biomass process streams and
for other improvements in process efficiencies, such as
oxygen enrichment of air gasifier intakes and alcohol-
water separations. This research includes (1) the
engineering performance evaluation of membranes in
alcohol production systems; (2) studies of mechanisms of
membrane fouling and degrading; and (3) development of
new membrane systems specifically designed for use in
alcohol production (Solar Energy Research Institute
1981).

Project: Removing Alcohol From Water by
Selective Absorption

Organization: Shock Hydrodynamics Division
                Whittaker Corp.
                4710-4716 Vineland Ave.
                North Hollywood, CA 91602

Telephone: 213/985-6940

Contact: Emil Lawton

Dates: 1981 - 1982

Synopsis: The objective of this project is to develop
an imbibitive polymer that will efficiently remove
alcohol from water by absorption. Imbibitive polymers
are usually prepared as tiny spheres and act as solid,
selective solvents. They have a high capacity as
absorbents rather than adsorbents, release their imbibed
liquids readily, and can be recycled repeatedly with
negligible loss of material. The use of such absorbents

has the advantages of both liquid-liquid extraction and column separations, but with considerably lower energy capital requirements.

An assessment of dollar and energy costs of the method will be made (Solar Energy Research Institute 1981).

Production Processes

Project: Biological Production of Liquid Fuels from Cellulose Biomass

Organization: University of Pennsylvania School of Medicine
Department of Biochemistry and Biophysics
Philadelphia, PA 19104

Telephone: 215/243-8038

Contact: E. Kendall Pye

Dates: 1976 - continuing

Synopsis: The objectives of this project are to investigate and solve the technical and economic problems relating to development of an integrated process for the total conversion of biomass (primarily poplar tree chips, but also other cellulose residues) into liquid fuels, chemical feedstocks, and other by-products. The hybrid poplar, clone 388, used for this research has several advantages. It grows well on marginal land, can be harvested in four years, and can be harvested year round. The major products are two liquid fuels: (1) ethanol for use as a gasoline extender, and (2) a butanol-lignin slurry for use as a bunker-C type heating fuel. Features of the process are the use of hot, aqueous butanol pretreatment to delignify the biomass and yield a solid phase, aqueous phase, and butanol phase. The solid phase comprises cellulose, which is highly degradable by cellulase enzymes; the aqueous phase comprises partially degraded hemicellulose, from which xylan for xylose production can be recovered; the butanol phase includes a high-quality polymer-grade lignin. The remaining lignin precipitates from the butanol as it cools to yield a butanol-lignin slurry, for use as a fuel. Elevated temperature fermentations are possible because of the use of Thermomonospora (earlier identified as belonging to the genus Thermoactinomyces), which produces a cellulase having extracellular and exo-glucanases of excellent thermal stability.

Specific research activities in relation to this project include examining the cellulolytic activities

produced when Thermomonospora is grown on
microcrystalline cellulose (Avicel). The CMCase and
Avicelase activities are extracellular, although the
β-glucosidase is cell-associated. Studies have been
conducted on enzymatic activities relating to time,
temperature, and pH.

The pretreatment and hydrolysis can produce a
glucose syrup of sufficient concentration to be used in
existing starch and sugar fermentation facilities. This
would permit an easy transition from grain feedstocks to
potentially more cost-effective lignocellulose feed
materials (Pye 1979; Hagerdal 1980).

Project: Cellulose-to-Ethanol Process Development

Organization: Georgia Institute of Technology
               Department of Chemical Engineering
               Atlanta, GA 30332

Telephone: 404/894-3530

Contact: Daniel O'Neil

Dates: 1980 - 1981

Synopsis: The work to date has utilized laboratory-
scale units. The technical feasibility of the
conversion processes will be evaluated, and scale-up and
cost factors will be estimated. The purpose of this
experimental program is to determine the technical and
economic feasibility of using fermentation to convert
cellulose biomass to ethanol. The process development
unit design will include (1) pretreatment of the
cellulose biomass to separate the cellulose from
hemicellulose and lignin; (2) acid hydrolysis of the
cellulose to glucose or other simple sugars;
(3) fermentation of the sugar to ethanol; and (4)
recovery of the ethanol (Solar Energy Research Institute
1981).

Project: Chemicals from Cellulose

Organization: University of Arkansas
               Biomass Research Center
               415 Administration Building
               Fayetteville, AK 72701

Telephone: 501/575-2251; 2654

Contact: George H. Emert

Dates: 1974 - continuing

Synopsis: This work was originally funded by the Gulf
Oil Corp. and was then taken over by the U.S. Department
of Energy. The pilot plant has been operating since
1976, and projections include at least partial
commercialization by the early 1980s. Feasibility of
constructing a plant to produce 25 million gal/yr of
ethanol is under study.

    A mutant strain of Trichoderma reesei--a mold found
in the soil--is being used. An essential step in the
process is to make the T. reesei produce the enzymes
needed to break down the cellulose into glucose in
sufficient quantity to sustain a large-scale operation.
Water is added to waste cellulose in a pulper. This
makes a slurry and breaks up the fibers, creating a
greater surface area and allowing the T. reesei access
to the cellulose. A portion of the pulp is sterilized
and fed into enzyme production tanks, where it is mixed
with nutrients and air and inoculated with the fungus.
The remaining pulp from the blender moves to the reactor
tanks and is mixed with the enzymes along with yeast and
nutrients in a controlled environment, where
simultaneous saccharification and fermentation take
place. The enzymes break down the cellulose to form
glucose (saccharification), and as soon as the molecules
of glucose form, the yeast catalyzes their conversion to
ethanol (fermentation) (Emert 1980).

Project: Design, Fabrication, and Operation of a
Biomass Fermentation Facility

Organization: Georgia Institute of Technology
                Engineering Experiment Station
                Atlanta, GA   30332

Telephone: 404/894-2865; 2889; 2882; 3411; 3655

Contact: Ronnie Roberts
          Alton Colcord
          John Muzzy

Dates: 1978 - continuing

Synopsis: Phase I of this study has resulted in a
conceptual design, a detailed engineering design, and a
cost estimate for construction of a 3-oven-dried
tons/day process development unit in order to assess the
technical and economic feasibility of ethanol production
from lignocellulose biomass (for example, softwoods,
hardwoods, corn stover, wheat straw). The design has
focused on a mainline process called the GIT process,

which is a dilute-acid hydrolysis process. The GIT
process is based on a unique combination of existing
technologies that have been demonstrated commercially.
This process was designed following a detailed survey of
acid and enzymatic processes that have been or are being
developed.

The GIT process utilizes a pretreatment by
decompression (steam explosion), and lignin extraction
by solvents at low temperatures and pressures.
Cellulose is separated from the lignin and
hemicellulose, and is converted in a fixed-bed reactor
to as much as 85% sugar. The hemicellulose remaining
after decompression is converted in the conventional
prehydrolysis stage (Solar Energy Research Institute
1980b).

Project: Ethanol Production Via Fungal Decomposition
and Fermentation of Biomass

Organization: Argonne National Laboratories
                9700 South Cass Ave.
                Argonne, IL 60439

Telephone: 312/972-2000; 3368

Contact: Antonios A. Antonopoulos

Dates: January 1981 - continuing

Synopsis: This project is testing a potentially
cost-effective, high-yield process for conversion of
lignocellulose biomass to ethanol. The microbiologial
system will provide all required operations for the
process from lignocellulose biomass to alcohol
production. The proposed Fusarium fungal system will
simultaneously degrade the lignin fraction of the
biomass and convert the cellulose and hemicellulose to
the respective 6- and 5-carbon sugars (glucose and
xylose). Fusarium has the ability to convert both
sugars to ethanol and thereby increase the net yield of
product/lb of substrate. The product yield should be
increased by between 20% - 50%, depending upon choice of
feedstock (Solar Energy Research Institute 1980b).

Project: Fermentable Sugars and Fermentation Products
from Cellulose Materials

Organization: Purdue University
                Laboratory of Renewable Resources
                Engineering
                A.A. Potter Engineering Center
                West Lafayette, IN 47907

Telephone: 317/749-6306

Contact: Cheng-shung Gong
           Christine M. Maun
           George T. Tsao

Dates: 1974 - continuing

Synopsis: Several research projects on fermentation are
being conducted at Purdue.   One study is examining the
production of ethanol from D-xylose by using D-xylose
isomerase and yeasts.   The results indicate that ethanol
can be produced from D-xylose in yields >80%.   First,
D-xylose is converted to D-xylulose by xylose isomerase,
then D-xylulose is fermented to ethanol by yeasts.   A
mutant strain of Candida sp.--XF217--produces ethanol
from D-xylose both aerobically and anaerobically.
Aerobic conditions produced a higher yield.   Work on
Monilia sp. indicates that it is a potential
microorganism for conversion of cellulose biomass to
ethanol (Gong   1981a).

Project: High-Temperature/High-Yield Hydrolysis of
Lignocellulose for Ethanol Production

Organization: Stake Technology, Ltd.
                220 Wyecroft Rd.
                Oakville, Ontario, L6K 3V1 Canada

Telephone: 416/842-4560

Contact: John Taylor

Dates: 1976 - continuiung

Synopsis: A continuous process for the high-pressure
steam treatment of lingnocelluose material has been
developed.   The system is in commercial operation in the
United States producing high-energy ruminant feed from
sugarcane bagasse and hardwood.
     Development work is continuing on a process for the
conversion of waste biomass into ethanol and coproducts.
The lignocellulose materials are separated into their
three major components--cellulose, hemicellulose, and
lignin--by aqueous and alkaline extractions.   The
cellulose fraction is hydrolized to glucose for
subsequent fermentation to ethanol.   The pentosan-rich
hemicellulose fraction can be converted to a variety of
chemicals, most notably furfural.   The lignin fraction
has been shown to have potential as an extender for
phenol-formaldehyde resins and to possess unique
pharmaceutical properties (Taylor   1981).

Project: Innovative Concepts in Application
of Organic Electrochemistry

Organization: University of California at Los Angeles
Engineering Center
3109 Murphy Hall
405 Hilgard Ave.
West Los Angeles, CA 90024

Synopsis: This research focuses on electrochemical
systems that offer better utilization of by-product
streams from cellulose-to-ethanol processes. The
applicability of preparative electroreduction as an
experimental method for producing high-energy fuels and
chemicals will be determined. The initial thrust of the
program will be the reduction of fermentation sugars and
polyalcohols to hydrocarbons (Solar Energy Research
Institute 1981).

Project: Liquid Fuel Production from Biomass

Organization: Dynatech Co.
99 Erie St.
Cambridge, MA 02139

Telephone: 617/868-8050

Contact: Donald L. Wise

Dates: Continuing

Synposis: This program is developing a basic process
for the production of liquid fuel from biomass. It
consists of three steps: (1) carboxylic acids are
produced from cellulose by nonsterile anaerobic
fermentation; (2) acids are separated and concentrated
by liquid-liquid extraction, and (3) the concentration
is converted by electrolytic oxidation (Kolbe
electrolysis) to the final alcohol fuel (Solar Energy
Research Institute 1980b).

Project: Liquid Fuels From Biomass Electrolysis

Organization: Institute of Gas Technology
3424 South State St.
IIT Center
Chicago, IL 60616

Telephone: 312/567-3650

Contact: Anthony F. Sammells

Dates: 1981 - 1982

Synposis: The objective of the program is to identify
electrosynthetic routes for the production of liquid
fuel from biomass-derived glucose.   In particular, the
electrochemical reduction of glucose dehydration
products in acid electrolyte--primarily
5-(Hydroxymethyl)-2-furfural--is expected to lead to the
liquid fuel 2, 5-dimethylfuran, which has combustion
characteristics similar to those of molecules found in
gasoline.   Electrosynthetic routes and electrolytic cell
designs for the direct conversion of biomass-derived
glucose to liquid fuel will be identified (Solar Energy
Research Institute 1981).

Project: Organosolv Delignification as a
Pretreatment for Enzymatic Hydrolysis

Organization: University of Washington
                Department of Forestry
                Seattle, WA 98195

Telephone: 206/543-1918

Contact: D.V. Sarkanen

Dates: 1981 - 1982

Synopsis: The purpose of this project is to develop
data for the design and operation of a pilot plant
facility for the separation of various biomass materials
into enzymatically hydrolyzable fibers, solid lignin
fuels and feedstocks, and sugars derived from the
hemicellulose components.   Extraction of the lignin into
alcohol-water mixtures is the approach to be
investigated.   Representative examples of three classes
of feedstocks will be studied:
  1.   hardwoods--namely, cottonwood, southern oak
       and sweetgum;
  2.   agricultural residues--namely, wheat and rice
       straws, and corn stover; and
  3.   bagasse, from sugarcane and sweet sorghum
       (Solar Energy Research Institute 1981).

ETHANOL PRODUCTION: ENERGY REQUIREMENTS AND ECONOMICS

Project: Ethanol Purification by Adsorption

Organization: Iowa State University
                Ames Laboratory
                Ames, IA 50011

Telephone: 515/294-3483

Contact: C.D. Chriswell
           R.W. Fisher

Dates: 1980 - 1981

Synopsis: An adsorption process under development has
the potential for reducing the energy required to
produce absolute or near-absolute ethanol from
fermentation beer and for simultaneously reducing the
cost of ethanol production facilities.   This process
uses a unique adsorbent capable of adsorbing ethanol
from dilute aqueous solutions and from which the ethanol
can be recovered using an energy efficient desorption
procedure.   Ethanol adsorption is a continuous process
that does not affect sugars or starches.   It is,
therefore, feasible to couple the adsorption
purification process with a continuously operating
fermentor and remove ethanol from beer at the same rate
it is formed.   This would reduce the size and cost of
fermentation equipment and decrease the energy required
for fermentation.
        The design parameters for the uptake and stripping
of alcohol on and from the adsorbent, and the
physiochemical properties of the adsorbent are being
examined to develop isothermal data (Solar Energy
Research Institute   1981).

Project: Fuel Alcohol Production by an Operating Farm-
Scale Plant: A Cost and Energy Study

Organization: South Dakota State University
                Department of Microbiology
                Brookings, SD   57006

Telephone: 605/688-4149

Contact: Raymond Moore

Dates: 1977 - continuing

Synopsis: This project is studying a pilot-scale
alcohol fuel plant.   The primary interests are in
fermentation; distillation; and use of the product mash,
or stillage.   The mash is being used in animal feeding
troughs and the alcohol is being tested in tractors and
farm pickup trucks.   The energy balance and economic
feasibility of small plants is also being examined
(Moore   1981).

Project: Sweet Sorghum Optimization Studies
for Fuel Alcohol Production

Organization: Holly Sugar Corp.
P.O. Box 1052
Colorado Springs, CO 80901

Telephone: 303/471-0123

Contact: Donald Dickenson

Dates: 1981 - 1982

Synopsis: This project is intended to add to alcohol production capacity by optimizing the addition of a sweet sorghum crop to the raw material base of beet sugar processing plants in the Imperial Valley in California. This addition would extend the operation time of the plants by an additional three months, for a total of at least six months each year. About half the land in the Imperial Valley stands idle during the optimal growing season for sweet sorghum, so no crops would be displaced.
Research will be done in the following areas:
1.   optimal date of planting and date of harvest;
2.   growing and harvesting patterns;
3.   size of cane for most efficient diffusion process; and
4.   juice purification and fermentation (Solar Energy Research Institute 1981).

ETHANOL PRODUCTION: IMPROVED TECHNIQUES USING SUGAR AND STARCH FEEDSTOCKS

Project: Continuous On-Farm Corn-Alcohol Production at 12 gal/day: Energy and Management Needs

Organization: Iowa State University
Agriculture and Economics Experiment Station
Ames, IA 50010

Telephone: 515/294-3917

Contact: Wesley F. Buchele

Dates: 1980 - 1982

Synopsis: The purpose of this project is to develop a 12 gal/day, automatically controlled alcohol fuel facility. It will utilize a continuous fermentation distillation cycle (Solar Energy Research Institute 1981).

Project: Energy Efficient Production of Anhydrous Alcohol

Organization: Purdue University
                Laboratory of Renewable Resources
                Engineering
                A.A. Potter Engineering Center
                West Lafayette, IN 47907

Telephone: 317/749-3593

Contact: Michael R. Ladisch
           Karen Dyck
           Juan Hong

Dates: 1980 - 1982

Synopsis: The purpose of this research is to develop a method of ethanol dehydration that decreases the amount of process energy. Various biomass materials, e.g., cornmeal, cornstarch, shelled corn, corn residues, and bagasse, were tested as drying agents. The principal of drying ethanol by a nondistillation process has historical precedent.
Cornmeal has been found to be quite stable and appears amenable to cyclic operation. The conditions best suited seem to be close to the dewpoint for adsorption and at $100° - 120°C$ ($212° - 248°F$) for regeneration. Air at 30% - 50% relative humidity seems to be appropriate for regeneration. The pressure drop in an adsorber packed with 20-mesh cornmeal is sufficiently low that it does not interfere with the distillation when directly connected to a distillation column. Runs with a pilot unit combining distillation and adsorption show no major practical operational difficulties. Estimates based on laboratory data and runs with the pilot unit show 11,500 Btu are required to obtain a 99.6% ethanol product starting from a 10.5% starting feed, and 13,000 Btu are required if the feed is 7.5% ethanol (Ladisch 1980).

Project: Liquid-Liquid Extraction Alcohol Separation

Organization: Georgia Institute of Technology
                Department of Chemical Engineering
                Atlanta, GA 30332

Telephone: 404/894-2856

Contact: D. William Tedder

Dates: 1981 - continuing

Synopsis: A low-energy liquid-liquid extraction process for separating alcohol from water is being developed (Douglas 1981a).

Project: Molecular-Sieve Alcohol Separation

Organization: Iowa State University
Ames Laboratory
Ames, IA 50012

Telephone: 515/294-4460

Contact: Ray W. Fisher

Dates: 1981 - continuing

Synopsis: This research is concentrating on alcohol separation techniques using hydrophobic molecular sieves that will preferentially adsorb alcohol from fermented beer (Douglas 1981a).

Project: Energy Efficient Water-Ethanol Separation Process

Organization: Hydrocarbon Research, Inc.
P.O. Box 6047
Lawrenceville, NJ   08648

Telephone: 609/394-3101

Contact: James Chao

Dates: 1981 - 1982

Synopsis: This project involves an energy efficient, two-adsorbent, two-stage water-ethanol separation process. It is based on using activated carbon and zeolite in a molecular sieve as adsorbents. The total energy consumed to dehydrate one gallon of ethanol starting from fermentation beer is expected to be about 7,220 Btu. This can be further improved by replacing the molecular sieve with an adsorbent that binds water less strongly (Solar Energy Research Institute 1981).

Project: Precipitation of Ethanol Stillage to Improve
Protein and Recycle Water

Organization: University of Nebraska
               Agriculture Experiment Station
               Lincoln, NE 68132

Telephone: 402/472-6443

Contact: Terry J. Klopfenstein

Dates: 1980 - 1983

Synopsis: This project is seeking a more efficient
method for precipitating the protein in stillage out of
water. This research is also attempting to recycle
water back through the system. Treatments are
promising, but are still in the laboratory phase
(Klopfenstein 1981).

METHANOL PRODUCTION

Project: Conversion of Biomass to Methanol

Organization: Science Applications, Inc.
               1710 Goodridge Drive
               McLean, VA 22102

Telephone: 703/821-4563

Contact: Edward I. Wan

Dates: 1978 - continuing

Synopsis: This work began with the identification and
development of alternative systems for producing
methanol from biomass resources. An empirical systems
model is being developed to demonstrate methanol-
from-biomass production systems so that parametric
analysis and sensitivity studies can be done. Based on
these analyses, critical paths and plant-siting criteria
and logistics will be determined. Recommendations for
technology development for methanol production will be
made at the completion of this project.
     To date, results of this study have indicated
that economic production of methanol from biomass is
largely dependent on the gasification process and
waste-heat utilization. High-temperature, low-pressure
gasification shows a considerable advantage over the
low-temperature, low-pressure gasification process for

methanol synthesis. The necessary syngas modification step also lowers the methanol conversion efficiency. To achieve higher conversion efficiency and competitive economics, a hybrid biomass-methane feedstock would be preferable. This project will also study biomass conversion to ethanol and high-alcohol fuel through thermochemical processes (Wan 1981).

Project: Development and Test Program for Methanol or Ammonia-Producing Gasifier

Organization: Solar Energy Research Institute
1617 Cole Blvd.
Golden, CO 80401

Telephone: 303/231-1000

Contact: Thomas Reed

Dates: 1980 - continuing

Synopsis: This program is concerned with the development and testing of a gasifier particularly suited to methanol and ammonia production. The unit is capable of producing a clean synthesis gas under pressure, which can be used directly or with minimal additional compression in a methanol synthesis process. Capacity is approximately 1 ton/day of biomass (Solar Energy Research Institute 1981).

Project: Syngas Catalysis for Alcohol Fuels

Organization: Solar Energy Research Institute
1617 Cole Blvd.
Golden, CO 80401

Telephone: 303/231-1000

Contact: James Smart

Dates: September 1981 - continuing

Synopsis: Synthesis gas ($CO + H_2$) is the starting material for production of methanol, mixed low-molecular-weight alcohol, synthetic gasoline, and heavier liquid fuels. Syngas is available from renewable resources by the oxygen gasification of wood. This research is concerned with providing an up-to-date review of the technology, and exploring homogeneous, synthesis gas catalysis. Conventional catalysis is heterogeneous. Recent advances in organometallic chemistry and techniques to couple such catalysts to solid supports offer some advantages (Solar Energy Research Institute 1981).

BUTANOL PRODUCTION

Project: Degradation of Cellulose Biomass and Its Subsequent Utilization for the Production of Chemical Feedstocks

Organization: Massachusetts Institute of Technology
               Department of Nutrition and Food Science
               Cambridge, MA 02139

Telephone: 617/253-2126; 3108

Contact: Daniel I.C. Wang
          Charles L. Cooney

Dates: Continuing

Synopsis: The goal is to extend the results of the previously developed MIT process toward development of a biomass-to-butanol process. The successful completion of the proposed research will (1) increase the production of sugars from biomass, which will improve the ethanol and butanol process economics; and (2) remove technical obstacles to developing an economically viable butanol process, including mutant selection and improved bioreactor design (Solar Energy Research Institute 1980b).

Project: Fuel Alcohol Extraction Technology Commercialization

Organization: Oak Ridge National Laboratory
               Chemistry Division
               Mail Stop 4500 S-EST
               Oak Ridge, TN 37830

Telephone: 615/576-5454

Contact: A.L. Compere
          W.L. Griffith
          J.M. Googin

Dates: Continuing

Synopsis: The fualex, or fuel alcohol extraction process, uses a combination of hydrocarbon and surfactant to remove neutral solvents, such as butanol,

ethanol, isopropanol, and acetone, from aqueous
solution. The hydrocarbon extractants may be fuels,
such as gasoline, furnace oil, and diesel fuel.
Surfactant concentrations ranging from 1 - 10 gal/L and
hydrocarbon concentrations ranging from 0.01 - 1 L/L of
aqueous solution are being investigated. The fualex
process is being tested on solutions containing 5%
wt/vol total neutral solvents, since this is near
maximum for the fermentation product stream. The
neutral solvents are removed in the form of an emulsion
that is white to light blue in the visible range. The
emulsion has potential for direct use in fuels or as an
intermediate for obtaining purified solvent (Compere
1980).

Project: Potential Feedstocks for Acetane-Butanol Fermentation

Organization: Colorado State University
Department of Agricultural and Chemical
Engineering
Fort Collins, CO 80523

Telephone: 303/491-5252

Contact: Antonio R. Moreira

Dates: Continuing

Synopsis: This research is evaluating various
feedstocks for acetone-butanol fermentation. Feedstocks
being tested include molasses and whey. To be
competitive with production of butanol from
hydrocarbons, acetone must be credited as a coproduct.
A major obstacle to commercialization is the low level
of butanol produced. Research continues on improving
yields (Solar Energy Research Institute 1981).

Project: Evaluation of Substrates

Organization: Oak Ridge National Laboratory
Chemistry Division
Mail Stop 4500 S-EST
Oak Ridge, TN 37830

Telephone: 615/574-4970

Contact: Alicia L. Compere

Dates: Continuing

Synopsis: This research is evaluating the following
substrates for acetone-butanol fermentation (1)
pentoses, (2) hexoses, (3) disaccharides, and (4)
polysaccharides. The substrates are being tested at
concentrations ranging between 2.5% - 10% wt/vol. The
following microorganisms are being tested: (1)
Clostridium butylicum strains NRRL B592 and NRRL B593,
(2) Clostridium acetobutylicum strains NRRL B527 and
NRRL B3179, (3) Clostridium pasteurianum strain NRRL
B598, (4) a mixed culture of the Clostridia and
Klebsiella pneumoniae strain NRRL B427, and (5) a mixed
culture of number 4 with the addition of a yeast
isolated from kefir culture. Clostridium butylicum
strains NRRL B592 and NRRL B593 were able to ferment
effectively carbohydrates that occur in dairy and wood
wastes (Compere 1979).

Project: Liquid Fuels and Chemical Production from
Cellulose Biomass: Hemicellulose and Cellulose
Hydrolysis Recovery, and Pentose Utilization in a
Biomass Processing Complex

Organization: Auburn University
                Department of Chemical Engineering,
                Chemistry and Microbiology
                Auburn, AL 36830

Telephone: 205/826-4827

Contact: Robert Chambers   Lon Mathias
           Y.Y. Lee   Tim Placek
           Tom McCaskey   Don Vives

Dates: 1977 - continuing

Synopsis: Fermentation of xylose to produce butanol is
being examined. Clostridium butylicum has provided the
highest yield of butanol of the 21 species studied.
Alternatives to the more expensive components of the
process were found; e.g., the nitrogen source originally
specified as 2% yeast extract can be successfully cut to
1%, and a less expensive inorganic nitrogen source can
be substituted for the yeast extract. Also, reduced
iron can be used to ensure anaerobic conditions instead
of the more costly sodium thioglycollate and L-cystine.
Test results indicate that inhibition is caused by the
butanol product rather than the substrate. Culture
adaptation against butanol appears necessary to make
this fermentation feasible (Solar Energy Research
Institute 1980b).

Project: The Effect of Ethanol and Butanol on Sugar
Transport and the Viability of Clostridium Bacteria

Organization: Colorado State University
               Department of Agricultural and Chemical
               Engineering
               Fort Collins, CO 80523

Telephone: 303/941-8288

Contact: James Linden

Dates: 1981 - continuing

Synopsis: The purpose of this project is to examine the
alcohol toxicity problem present in acetone-butanol
fermentation. In this process, the total concentration
of the neutral solvents is typically 2% - 3%, the point
at which solvent levels are toxic to Clostridium.
Results suggest that simple modifications to cell growth
or fermentation procedures could suppress alcohol
toxicity.
    The same concentrations of ethanol and butanol that
are toxic to Clostridium also affect energy-driven sugar
transport mechanisms in a variety of cells. Since the
sugar transport process is so closely tied to cell
viability, it is possible that toxic levels of alcohol
affect cell viability by disrupting the sugar transport
mechanism located in the cell membrane of Clostridium
organisms. The objectives of this program are to:
   1. determine the mechanisms of sugar transport in
       Clostridium species;
   2. determine the effect of low concentrations of
       ethanol and butanol on sugar transport in
       these organisms;
   3. establish the threshold concentrations of
       ethanol and butanol that affect the viability
       of these organisms;
   4. evaluate membrane composition differences in
       Clostridium species that may stabilize the
       sugar transport system and cell viability in
       the presence of alcohol;
   5. attempt growth on varying membrane composi-
       tions to establish a stable sugar transport
       system; and
   6. study the yeast-sugar transport system so that
       higher alcohol concentrations might be
       possible (Solar Energy Research Institute
       1980b).

Project: Biological Production of Liquid Fuels from
Cellulose Biomass

Organization: University of Pennsylvania School of
               Medicine
               Department of Biochemistry and Biophysics
               Philadelphia, PA 19104

Telephone: 215/243-8038

Contact: E. Kendall Pye

Dates: 1976 - continuing

Synopsis: The objectives of this project are to
investigate and solve the technical and economic
problems relating to development of an integrated
process for the total conversion of biomass (primarily
poplar tree chips, but also other cellulose residues)
into liquid fuels, chemical feedstocks, and other
by-products. The hybrid poplar, clone 388, used for
this research has several advantages. It grows well on
marginal land, can be harvested in four years, and can
be harvested year round. The major products are two
liquid fuels: (1) ethanol for use as a gasoline
extender, and (2) a butanol-lignin slurry for use as a
bunker-C type heating fuel. Features of the process are
the use of hot, aqueous butanol pretreatment to
delignify the biomass and yield a solid phase, aqueous
phase, and butanol phase. The solid phase comprises
cellulose, which is highly degradable by cellulase
enzymes; the aqueous phase comprises partially degraded
hemicellulose, from which xylan for xylose production
can be recovered; the butanol phase includes a
high-quality polymer-grade lignin. The remaining lignin
precipitates from the butanol as it cools to yield a
butanol-lignin slurry, for use as a fuel. Elevated
temperature fermentations are possible because of the
use of Thermomonospora (earlier identified as belonging
to the genus Thermoactinomyces), which produces a
cellulase having extracellular and exo-glucanases of
excellent thermal stability.
     Specific research activities in relation to this
project are examining the cellulolytic activities
produced when Thermomonospora is grown on
microcrystalline cellulose (Avicel). The CMCase and
Avicelase activities are extracellular, although the
β-glucosidase is cell-associated. Studies have been
conducted on enzymatic activities relating to time,
temperature, and pH.

The pretreatment and hydrolysis work can produce a glucose syrup of sufficient concentration to be used in existing starch and sugar fermentation facilities. This would permit an easy transition from grain feedstocks to potentially more cost-effective lignocellulose feed materials (Pye 1979; Hagerdal 1980).

REFERENCES

Baker, Andrew J. 1980. "Gasohol from Wood is Not Yet
Economically Feasible." Forest Farmer. Vol. XL, no.
2. Madison, WI: Forest Products Laboratory, USDA.

Blanch, Harvey. 1981. University of California at
Berkeley, Berkeley, CA. Personal communication.

Clement, L. Davis. 1981. Texas Tech University,
Department of Chemical Engineering, Lubbock, TX.
Personal communication.

Compere, A.L., Griffith, W.L. 1979. "Evaluation of
Substrates for Butanol Production." Development in
Industrial Microbiology. Vol. 20. Society for
Industrial Microbiology.

Compere, A. L., Griffith, W. L., Googin, J. M. 1980.
"Fuel Alcohol Extraction Technology Commercialization
Conference." Oak Ridge, TN: Oak Ridge National
Laboratory, Chemistry Division.

Converse, Alvin O., Grethlein, Hans E. 1979. "Acid
Hydrolysis of Cellulosic Biomass." Hanover, NH:
Dartmouth College, Thayer School of Engineering.

Crawford, Don L. 1978. "Lignocellulose Decomposition
by Selected Streptomyces Strains." [In Applied and
Environmental Microbiology.] Moscow, ID: University of
Idaho, Department of Bacteriology and Biochemistry.

Crawford, Don L. 1979. "Bioconversion of Plant
Residues into Chemicals: Production of Chemicals from
Lignin." Moscow, ID: University of Idaho, Department
of Bacteriology and Biochemistry.

Crawford, Don L., Crawford, Ronald L. 1980. "Microbial
Degradation of Lignin." [In Enzyme Microbiology
Technology. Vol. 2.] Moscow, ID: University of Idaho,
Department of Bacteriology and Biochemistry.

Crawford, Don L., Phelen, Mary Beth, Pometto, Anthony
L. 1979. "Isolation of Lignocellulose-decomposing
actinomycetes and degradation of specifically

182

^{14}-C-labeled lignocelluloses by six selected
Streptomyces Strains." [In Canadian Journal of
Microbiology. Vol. 25, no. 11.] Moscow, ID:
University of Idaho, Department of Bacteriology and
Biochemistry.

Cuskey, S. M., et al. 1980. "Screening for
β-Glucosidase Mutants of Trichoderma reesei with
Resistance to End-Product Inhitibtion." Developments in
Industrial Microbiology. Vol. 21. Society for
Industrial Microbiology.

Dobbs, Thomas L. and Hutchinson, Ron. 1980.
"Preliminary Cost Estimates--Producting Alcohol Fuel
From a Small Scale Plant." Brookings, SD: South Dakota
State University, Department of Microbiology.

Douglas, Larry. 1981a. Solar Energy Research
Institute, Golden, CO. Personal communication.

Douglas, Larry. 1981b. The Chemistry and Energetics of
Biomass Conversion, an Overview. Golden, CO: Solar
Energy Institute.

Emert, G. H., et al. 1980. "Economic Update of the
Bioconversion of Cellulose to Ethanol." Chemical
Engineering Progress. Vol. 76, no. 9, pp. 47-52.

Eveleigh, D.E., Montenecourt, B.S., Kelleher, T.J.
1980. "Biochemical Nature of Cellulases from Mutants of
Trichoderma reesei." Biotechnology and Bioengineering.
Symposium. No. 10, pp. 15-26. New York, NY: John
Wiley & Sons, Inc.

Ferchak, John D., Pye, E. Kendall. 1981. "Utilization
of Biomass in the U.S. for the Production of Ethanol
Fuel as a Gasoline Replacement." Solar Energy. Vol 26,
pp. 9-16. Elmsford, NY: Pergamon Press, Ltd.

Glasser, Wolfgang G. 1981. Virginia Polytechnic
Institute and State University, Blacksburg, VA.
Personal communication.

Glasser, Wolfgang G., Drew, Stephen W., Hall, Philip L.
1980. "Enzymatic Transformations of Lignin."
Blacksburg, VA: Virginia Polytechnic Institute and
State University.

Goldstein, Irving S. 1980. "An Integrated Approach to
the Conversion of Lignocellulose from Wood into Useful
Chemicals." Raleigh, NC: North Carolina State
University, Department of Wood and Paper Science.

Gong, Cheng-shung, McCracken, Linda D., Tsao, George T. 1981a. "Direct Fermentation of D-Xylose to Ethanol by a Xylose-Fermentating Yeast Mutant, Candida sp. XF 217." West Lafayette, IN: Purdue University, Laboratory of Renewable Resources Engineering.

Gong, Cheng-shung, Maun, Christine M., Tsao, George T. 1981b. "Direct Fermentation of Cellulose to Ethanol by a Cellulolytic Filamentous Fungus." West Lafayette, IN: Purdue University, Laboratory of Renewable Resources Engineering.

Gong, Cheng-shung, et al. 1981c. "Production of Ethanol from D-Xylose by Using D-Xylose Isomerase and Yeasts." West Lafayette, IN: Purdue University, Laboratory of Renewable Resources Engineering.

Grethlein, Hans E. 1978. "Comparison of the Economics of Acid and Enzymatic Hydrolysis of Newsprint." Biotechnology and Bioengineering. Vol. XX, pp. 503-525. New York, NY: John Wiley & Sons, Inc.

Grohmann, Karl. 1981. Solar Energy Research Institute, Golden, CO. Personal communication.

Hagerdal, Barbel, Ferchak, John D., Pye, E. Kendall. 1980. "Saccharification of Cellulose by the Cellulolytic Enzyme System of Thermomonospora sp." Biotechnology and Bioengineering. Vol. XXII, pp. 1515-1526. New York, NY: John Wiley & Sons, Inc.

Harris, Elwin E., Beglinger, Edward. 1946. "The Madison Wood-Sugar Process." Madison, WI: Forest Products Laboratory, USDA.

Harris, John F. 1975. "Acid Hydrolysis and Dehydration Reactions for Utilizing Plant Carbohydrates." Applied Polymer Symposium. No. 28, pp. 131-144. New York, NY: John Wiley & Sons, Inc.

Jet Propulsion Laboratory. 1978. "Bioconversion Study Conducted by JPL." Pasadena, CA: California Institute of Technology.

Kelsey, Rick G., Shafizadeh, Fred. 1980. "Enhancement of Cellulose Accessibility and Enzymatic Hydrolysis by Simultaneous Wet Milling." Biotechnology and Bioengineering. Vol. XXII, pp. 1025-1036. New York, NY: John Wiley & Sons, Inc.

Klopfenstein, Terry J. 1981. University of Nebraska, Agriculture Experiment Station, Lincoln, NE. Personal communication.

Knappert, Diane, Grethlein, Hans, Converse, Alvin.
1980. "Partial Acid Hydrolysis of Cellulosic Materials
as a Pretreatment for Enzymatic Hydrolysis."
Biotechnology and Bioengineering. Vol. XXII, pp.
1449-1463. New York, NY: John Wiley & Sons, Inc.

Ladisch, Michael R., Dyck, Karen. 1979. "Dehydration
of Ethanol: New Approach Gives Positive Energy
Balance." Science. Vol. 205, no. 4409, pp. 898-900.
Washington, D.C.: American Association for the
Advancement of Science.

Ladisch, Michael R., Hong, Juan. 1980. "Energy
Efficient Production of Anhydrous Alcohol." Washington,
D.C.: Science and Education Administration, Cooperative
Research, USDA.

Ljungdahl, Lars. 1981. University of Georgia,
Department of Biochemistry, Athens, GA. Personal
communication.

Lucas, Linda E. 1981. NOVO Laboratories Inc., Wilton,
CT. Personal communication.

Mandels, Mary. 1981. U.S. Army, Natick R & D
Laboratories, Natick, MA. Personal communication.

Millett, Merrill A., Baker, Andrew J., Satter, Larry D.
1976. "Physical and Chemical Pretreatments for
Enhancing Cellulose Saccharification." Biotechnology
and Bioengineering Symposium. No. 6, pp. 125-153. New
York, NY: John Wiley & Sons, Inc.

Moore, Raymond. 1981. South Dakota State University,
Department of Microbiology, Brookings, SD. Personal
communication.

NOVO. n.d. "Application of Novo Enzymes for Fuel
Ethanol Production." Wilton, CT: NOVO Laboratories,
Inc.

NOVO. n.d. "Celluclast." Wilton, CT: NOVO
Laboratories, Inc.

Peterson, Gene. 1981. California Institute of
Technology, Jet Propolsion Laboratory, Pasadena, CA.
Personal communication.

Pye, E. Kendall, Humphrey, Arthur E. "Production of
Liquid Fuels from Cellulosic Biomass." Presented at
Third Annual Biomass Energy Systems Conference, 5-7
June, 1979. Golden, CO: Solar Energy Research
Institute.

Reed, Thomas. 1981. Solar Energy Research Institute, Golden, CO. Personal communication.

Reilly, Peter J. 1979. "The Conversion of Agricultural By-Products to Sugars." Ames, IA: Iowa State University, Department of Chemical Engineering.

Rugg, Barry, Brenner, Walter. 1980a. "Continuous Acid Hydrolysis of Waste Cellulose for Ethanol Production." New York, NY: New York University, Department of Applied Science.

Rugg, Barry, Armstrong, Peter, Stanton, Robert. 1980b. "Preparation of Glucose Substrates for Fermentation from Cellulosics Using the NYU Continuous Acid Hydrolysis Process." New York, NY: New York University, Department of Applied Science.

Rugg, Barry, Armstrong, Peter, Stanton, Robert. 1980c. "The New York University Continuous Acid Hydrolysis Process-Hemicellulose Utilization - Preliminary Data and Economics for Ethanol Production." New York, NY: New York University, Department of Applied Science.

Ryu, Dewey D.Y., Mandels, Mary. 1980. "Cellulases: Biosynthesis and Applications." Enzyme Microbiology Technology. Vol. 2. IPC Business Press.

Saeman, Jerome. 1981. USDA, Forest Service, Madison, WI. Personal communication.

Solar Energy Research Institute. 1980a. Biomass Refining Newsletter. Golden, CO: Solar Energy Research Institute.

Solar Energy Research Institute. 1980b. Alcohol Fuels Process R/D Newsletter. Golden, CO: Solar Energy Research Institute.

Solar Energy Research Institute. 1981. "Alcohol Fuels Activities at the Solar Energy Research Institute." Golden, CO: Solar Energy Research Institute.

Spano, Leo A., Mandels, Mary. 1979. "Enzymatic Hydrolysis of Cellulose to Fermentable Sugar for Production of Ethanol." Natick, MA: U.S. Army Natick Research and Development Command.

Stevenson, Tom. 1981. University of Montana, Department of Chemistry, Missoula, MT. Personal communication.

Sundstrom, D. W., et al. 1981. "Enzymatic Hydrolysis
of Cellulose to Glucose Using Immobilized
β-Glucosidase." Biotechnology and Bioengineering.
Vol. XXIII, pp. 473-485. New York, NY: John Wiley &
Sons, Inc.

Taylor, John. 1981. Stake Technology, Ltd., Oakville,
Ontario, Canada. Personal communication.

Thompson, David R., Grethlein, Hans E. 1979. "Design
and Evaluation of a Plug Flow Reactor for Acid
Hydrolysis of Cellulose." I&EC Product Research
Development. Vol. 18, p. 166. Washington, D.C.:
American Chemical Society.

Tuse, D. Chou, T.W., Mason, B.J., Skinner, W.A. n.d.
"Activity Profiles of the Thermostable Cellulase of
Thielavia Therrestris." Menlo Park, CA: SRI (Stanford
Research Institute) International.

Villet, Ruxton. 1980, 1981. Solar Energy Research
Institute, Golden, CO. Personal communication.

Wan, Edward I. 1981. Science Applications, Inc.,
McLean, VA. Personal communication.

Wayman, Morris. 1977. "New Opportunities for Fuel From
Biological Processes." Toronto, Canada: University of
Toronto, Department of Chemical Engineering.

Wayman, Morris. 1981. University of Toronto,
Department of Chemial Engineering, Toronto, Canada.
Personal communication.

Wilke, Charles R., Blanch, Harvey W. 1979. "Process
Development Studies on the Bioconversion of Cellulose
and Production of Ethanol." Berkeley, CA: University
of California.

Zabriskie, Dane W., Qutubuddin, A.S.M., Downing, Kim M.
1979. "Production of Ethanol From Cellulose Using a
Soluble Cellulose Derivative as an Intermediate."
Buffalo, NY: University of New York, Department of
Chemical Engineering.

Zeikus, J. G. 1981. University of Wisconsin at
Madison, Department of Bacteriology, Madison, WI.
Personal communication.

Zeikus, J. G., et al. 1981. "Thermophilic Ethanol Fermentations." Trends in the Biology of Fermentations for Fuels and Chemicals. A. Hollander, ed. New York, NY: Plenum Press.

Zerbe, John I. 1980. "Outlook for Chemical Wood." 1980 Agricultural Outlook. Washington, D.C.: U.S. Government Printing Office.

7
Legislation

INTRODUCTION

The change of federal government administration in January 1981 radically affected federal policy. Although policy focuses primarily on alcohol fuel production, these policy changes will also impact use. The Carter administration's policy appeared to be to get as much started as quickly as possible, in hopes that high production goals could be met. The Reagan administration appears to be less concerned with cutting back oil imports and more concerned with establishing and implementing political philosophy. At the core of that philosophy is a belief in the ability of the private sector and the marketplace to resolve technical development problems.

Feasibility studies and cooperative agreements, advocated by the Carter administration to support their policy of getting alcohol fuel plants built, have been rescinded by the Reagan administration. Funding for these programs had been available through the U.S. Department of Energy (DOE) and the U.S. Department of Agriculture's (USDA) Farmers Home Administration (FmHA). This policy change has included rescinding feasibility study and cooperative agreement programs for which funding had already been appropriated, and halting projects for which negotiations had been completed. Loan guarantee programs for which funding had been appropriated are being continued, but at reduced funding levels (Merritt 1981).

Fifteen commercial ethanol plants have been earmarked by DOE to receive loan guarantees. Their total capacity is planned at approximately 475 million gal/yr, with capital costs estimated at $1.15 billion. It was expected that the loan guarantee funds will be leveraged up to a 3:1 ratio. Funding of $525 million had been appropriated for this program in FY 1980. The Supplemental Appropriations and Recision Act of 1981 decreased the amount of this funding to $271 million (Moorer 1981). Funds for feasibility studies for 13

ethanol plants and cooperative agreements for six
ethanol plants, two of which reportedly had been
negotiated to settlement, will be dropped (<u>Alcohol Week</u>
1981). Methanol plant feasibility studies and a
negotiated cooperative agreement for one methanol plant
will also be cut. The 25 applications for loan
guarantees and direct loans for ethanol plant
development analyzed by the FmHA will be dropped.
Funding of $525 million for this program has been cut.
    Fifteen ethanol plants had been approved for loan
guarantees under funding appropriated in FY 1980 and
extended to FY 1981. These projects were frozen by the
new administration so that they could be reevaluated.
As of mid-1981 this reevaluation was almost complete,
and it is expected that the majority of these plants
will be approved. There is some concern in Congress,
however, that USDA will not approve these plants before
the end of FY 1981, therefore, Congress has appropriated
$250 million for FY 1982 to insure the funding of this
loan guarantee program (Merritt 1981).
    While the Reagan administration expected the
program funding cuts to spur private investment in
potential plants, it has had the opposite effect.
Federal reneging on prior commitments and general
uncertainty about future federal action effectively
slowed investment in the alcohol fuels industry, at
least temporarily, in early 1981.
    DOE's Office of Alcohol Fuels is responsible for
funding long-range research and development projects
with a FY 1982 budget of $10 million. It is possible
that the quasi-governmental Synthetic Fuels Corporation
will take over some responsibility for alcohol fuel
plant development. It is expected that much of this
research will be done at the Solar Energy Research
Institute in Golden, CO (Moorer 1981).
    The Reagan administration expects to continue tax
credits for alcohol fuels, which amount to 4¢/gal at the
pump. Officials expect that the removal of price
controls on domestic crude oil will aid the
competitiveness of alcohol fuels in the marketplace.

FEDERAL LEGISLATION: ALCOHOL FUEL PRODUCTION[1]

Rural Development Act of 1972

Status: Passed, Public Law 92-419.

---

[1]Federal legislation is listed consecutively by public
law number.

Summary: Provides loan guarantees for farm operating or ownership loans, which may be used for alcohol fueled equipment or alcohol fuel plants. Loans are available to operators of family farms who are not able to obtain credit elsewhere.

Food and Agriculture Act of 1977

Status: Passed, Public Law 95-113.
Summary: Provides $60 million in loan guarantees to build four pilot alcohol fuel plants in the United States.

Energy and Water Development Appropriations Act of 1980

Status: Passed, Public Law 96-69.
Summary: Appropriates $54 million for biomass research and development. (DOE estimates $25 million will go toward alcohol fuels.)

Agricultural Appropriation Act for FY 1980

Status: Passed, Public Law 96-108.
Summary: Provides $500 million limit for loan guarantees for alcohol fuel production pilot projects, and $1 million limit for alcohol fuel research.

Interior Department Appropriations For FY 1980

Status: Passed, Public Law 96-126.
Summary: Appropriates $19 billion for Energy Security Reserve Fund to stimulate domestic commercial production of alternative fuels. DOE is providing $2.2 billion from the reserve fund for development of synfuels, including alcohol fuels.

Crude Oil Windfall Profit Tax Act of 1980

Status: Passed, Public Law 96-223.
Summary: Alcohol fuel provisions: a tax credit of $3 is provided for the production of liquid fuel from biomass in an amount equivalent to the energy content in a barrel of crude oil; four-cent-per-gallon excise-tax exemptions are extended through 1995 for ethanol-gasoline blends with 10% alcohol (190 proof); users of straight alcohol fuel are eligible for tax credit of 40 cents per gallon of alcohol that is at least 190 proof, and 30 cents per gallon of alcohol that is at least 150 proof but not more than 190 proof; blenders of ethanol-gasoline blends are eligible for tax credit of 40 cents per gallon of alcohol that is at least 190 proof, and 30 cents per gallon of alcohol that is at least 150 proof but not more than 190 proof; tax credit

is reduced by the value of the excise-tax exemption on
ethanol-gasoline blends; alcohol produced from
petroleum, natural gas, or coal does not qualify for the
tax credits or the excise-tax exemptions; the U.S.
Department of the Treasury is required to present a
report to Congress within six months of passage of the
act on what methods, if any, can be used to limit
alcohol fuel imports; acceptable means for denaturing
alcohol for fuel are clarified; annual report is
required from DOE on alcohol fuels, including
state-by-state usage, revenue loss to Treasury from tax
incentives, and production and retail costs; the Bureau
of Alcohol, Tobacco and Firearms (BATF) regulations
governing alcohol fuel production are eased; the 10%
Energy Investment Tax Credit for alcohol fuel production
equipment is extended through 1985; tax exemptions may
be used to finance solid-waste disposal facilities that
convert solid waste into alcohol fuel.

Amendment to Agricultural Act of 1949

Status: Passed, Public Law 96-234.
Summary: Amends the Agricultural Act of 1949 to
authorize the sale of Commodity Credit Corporation
stocks of corn for use in making alcohol for
motor vehicle fuel.

The Biomass Energy and Alcohol Fuels Act of 1980

Status: Passed, Public Law 96-294.
Summary. Provides funding of $1.05 billion for loan
guarantees, price guarantees, and purchase agreements
for biomass and alcohol fuel projects. Funding is
appropriated to USDA and DOE.

Rural Development Policy and Coordination Act of 1979

Status: Passed, Public Law 96-355.
Summary: Amends the Rural Development Act of 1972;
directs the secretary of agriculture to guarantee loans
of as much as $180 million for pilot projects for the
production of alcohol fuels from agricultural
commodities and forest products; authorizes the
secretary to promote research and development efforts
related to appropriate technology for small and
moderate-sized family farms.

Low-Income Energy Assistance Appropriations for FY 1981

Status: Passed, Public Law 96-369.
Summary: Makes continuing appropriations for FY 1981;
makes appropriations for the low-income energy
assistance biomass-energy and alcohol-fuels program, and

reserves for loan guarantee programs; continues funding at FY 1980 level for emergency energy conservation services; restricts funding for the Solar Energy and Energy Conservation Bank; and makes appropriations for other purposes.

Agricultural Act of 1980

Status: Passed, Public Law 96-494.
Summary: Title II under the Agricultural Trade Suspension Act of 1980 authorizes USDA to establish an ethanol feedstock reserve of any agricultural commodity exports that are suspended or restricted for national security or foreign policy reasons; authorizes USDA to establish loan programs for alcohol fuel processors to allow them to purchase and store grain; and authorizes USDA action for other purposes.

Bankruptcy Tax Act of 1979

Status: Passed, Public Law 96-589.
Summary: Amends the Internal Revenue Code of 1954 relative to bankruptcy; credit for ethanol-gasoline blends is included as a tax attribute for purposes of computing gross income when taxpayer deducts indebtedness discharge from tax attributes.

Supplemental Appropriations and Recision Act of 1981

Status: Passed, Public Law 97-12.
Summary: Deletes and decreases funding for a variety of programs including the following alcohol fuels programs: (1) decreases substantially funding for experimental alcohol fuels plants through USDA, (2) decreases loan guarantees from $525 million to $271 million through DOE, (3) deletes funding for feasibility studies and cooperative agreements through USDA and DOE, (4) decreases funding for information services on alcohol fuels by budget cuts for DOE, Solar Energy Research Institute (SERI), and USDA.

STATE LEGISLATION: ALCOHOL FUEL PRODUCTION[1]

---

[1] State legislation may be reported as passed, which is the year it was passed by a vote of the legislature; signed, which is the year it was signed by the governor; or enacted, which is the year the legislation went or is scheduled to go into effect. State legislation is listed alphabetically by state and consecutively by year passed, signed, or enacted.

Arkansas

H.B. 29/S.B. 28. Passed 1980. Authorizes the state
energy office to maintain the state alcohol fuel
registry.

California

S.B. 1395. Enacted 1976. Requires the State Solid
Waste Management Board, after consultation with the
State Energy Resources Board, to determine the economic
feasibility of developing a research and demonstration
program for the conversion of agricultural wastes to
synthetic fuel.

S.B. 771. Enacted 1979. Directs the State Energy
Resources Conversion and Development Commission to
implement a program to demonstrate residue conversion
technologies; at least 20 project sites will be
established to convert residue into synthetic fuels;
appropriates $15 million for these purposes.

A.B. 3171. Passed 1980. Authorizes local agencies to
contract for facilities for the conversion of solid
waste into energy, synthetic fuels, or reusable
materials, and to finance such facilities with specified
revenue bonds or by other means.

Colorado

S.B. 80. Passed 1978. Creates a nine-member committee
to promote the production and use of Gasohol, alcohol,
and related industrial hydrocarbons from Colorado
agricultural and forest products. The bill appropriates
$8,000 for administration. The bill also provides a
production incentive through a property-tax reduction
for personal or real property used exclusively in
producing ethanol for ethanol-gasoline belnds if the
quantity of ethanol produced each year is 2.5 million
gallons or less. For those businesses meeting the above
production requirements, the following schedule for
property-tax reductions applies for the 1980-1984 tax
years:
>   1st year--2% assessment of actual personal- or
>   real-property value;
>   2nd year--9% assessment of actual personal- or
>   real-property value;
>   3rd year--16% assessment of actual personal- or
>   real-property value;

4th year--23% assessment of actual personal- or
real-property value;
Thereafter--30% assessment of actual personal- or
real-property value.

H.B. 1607. Signed 1979. Expands the definition of
Gasohol to include motor vehicle fuels containing
alcohol derived from hydrocarbon or carbon-containing
by-products or waste products; grants a reduction in the
property tax on facilities used for the production of
such alcohol; and authorizes a five-cent tax exemption
to apply to blends of gasoline and alcohol with a purity
of at least 95%, and which have been produced from
Colorado products derived from hydrocarbon or
carbon-containing by-products, waste products, or
agricultural forest products.

Florida

S.B. 903. Signed 1980. Exempts the sale and
distribution of certain motor vehicle fuels blended with
alcohol from the first seven cents of the gas tax for a
designated period; provides for ethanol fuel development
tax-incentive credit to be allowed against the tax
imposed under the Florida Income Tax Code to
corporations that establish a new business or expand an
existing business engaged in the production of ethanol
or manufacture of equipment for the processing and
distillation of ethanol for motor vehicle fuels;
authorizes the sale at retail service stations of
alcohol-blended fuels meeting certain specifications.

Hawaii

S.B. 1581. Passed 1979. Sets appropriations for
alternative-energy research and a plant capable of
producing 700,000 gal/yr of ethanol per year for
ethanol-gasoline belends. The act also appropriates
$330,000 to establish a corn-to-ethanol research and
development program.

Illinois

H.B. 3403. Passed 1980. Directs the Institute of
Natural Resources to implement research, provide
information, and hold public seminars to promote
increased production and use of alcohol fuels.

Iowa

1973 Iowa Acts, Chapter 130, Sections 2, 3; Iowa Code
Sections 28.4, 28.7. Enacted 1973. Amends the duties
of the Iowa Development Commission to include the
development of an industry based on grain-alcohol
motor vehicle fuel and related products.

H.F. 734. Enacted 1979. Appropriates $50,000 to the
Iowa Development Commission for FY 1979 for the
promotion of Gasohol and associated by-products; extends
the time period for funds appropriated for ethanol
research until the end of FY 1980; Iowa State University
of Science and Technology is conducting this research
project.

Kansas

H.B. 2345. Enacted 1979. Authorizes transfer of funds
totaling $60,000 from the Corn Commission, Soybean
Commission, and Wheat Commission to the Kansas Energy
Office for the purpose of studying and analyzing grains
to be used as energy-resource alternatives.

Kentucky

H.C.R. 29. Enacted 1980. Requests the governor and the
secretary of energy to study the feasibility of
converting Kentucky distilleries not presently in
operation to ethanol production for use in producing
ethanol-gasoline blends. This resolution was solely a
House initiative and, as such, was not acted upon by the
Senate or governor.

Louisiana

S.B. 99. Passed 1979. Requests that the Department of
Natural Resources conduct a feasibility study on
obtaining methane gas from sugarcane.

H.B. 1033. Enacted 1979. Provides for the sale of
and/or delegation of processing rights for oil obtained
as in-kind royalties from state-owned mineral rights to
refiners with facilities for the distillation of
methanol or ethanol. The alcohol fuels must be suitable
for blending with gasoline to produce a motor vehicle
fuel, and at least 50% of the alcohol is to be derived
from agricultural products produced in Louisiana.

Maryland

S.B. 823. Enacted 1979. Permits the Maryland Industrial
Development Financing Authority to encourage and insure
loans for development and production of ethanol for
blending with gasoline.

H.B. 423. Signed 1980. Creates the Alcohol Plant Fund
for the construction and operation of alcohol
manufacturing and distributing facilities, to be
administered by the Board of Public Works; alters the
motor-vehicle fuel taxes to finance the fund; and
encourages use of ethanol-gasoline blends. Amended to
strike all references to the Alcohol Plant Fund;
increases the current state tax reduction of one cent
per gallon less than other motor vehicle fuels for
Gasohol to four cents per gallon.

S.J.R. 7. Enacted 1980. Encourages the Congress to
pass legislation promoting the development, production,
and use of domestically produced agricultural and
biomass alcohol fuels in the United States; provides for
sending copies of the resolution to state legislative
leaders in Indiana, Illinois, Iowa, Kentucky, Michigan,
Minnesota, Ohio, and Wisconsin, as well as to Maryland
members of Congress.

Minnesota

Laws of Minnesota for 1977, Chapter 381. Enacted 1977.
Directs the Minnesota Energy Agency to contract with the
University of Minnesota's Departments of Agricultural
Engineering and Agricultural and Applied Economics to
conduct a research and demonstration project regarding
the feasibility of developing an agriculturally derived
ethanol supplement to be blended with diesel fuel for
use in diesel engines for agricultural purposes.
Project's report is to include: results from field
studies of demonstration projects; a review of the
technical feasibility, possible impact on energy
resources and use, biomass options, economic
feasibility, agricultural sources, and policy
recommendations regarding ethanol supplements for diesel
fuels. Law was part of an overall 1977 energy
conservation measure.

H.F. 1495. Enacted 1979. Provides a residential energy
credit of 20% of the first $10,000 of renewable energy
source expenditures; includes expenditures that qualify
for the federal renewable energy credit, earth-sheltered
dwelling units, passive solar systems, and biomass
conversion equipment that produces ethanol, methane, or

methanol not for resale; effective 31 December 1978 to 1 January 1983.

Laws of Minnesota for the 1979 Special Session, Chapter 2. Passed 1979. Makes a $50,000 appropriation to the Minnesota Energy Agency for technical and nontechnical literature, including ethanol literature, and for coordination of adult and postsecondary energy-education programs.

Mississippi

S.B. 2234. Signed 1980. Changes the definition of "alcoholic beverage" to exclude ethyl alcohol (ethanol) used solely for fuel purposes; provides that it shall not be unlawful to own or possess a distillery used exclusively for the distillation of ethanol for fuel purposes.

Montana

S.B. 520. Enacted 1979. Authorizes the Department of Agriculture to contract for research and development of fuels from wheat and barley.

S.B. 523. Passed 1979. Establishes a special property-tax classification for property used in the production of ethanol for blending with gasoline.

S.J. Res. 28. Passed 1979. Requests the Department of Natural Resources and Conservation to involve the citizenry of Montana in programs administered by the department and related to the research and development of fuels manufactured from agricultural products and residues.

Nebraska

L.B. 424. Passed 1978. Provides matching funds (as much as $500,000) to any city, county, or village wishing to build an plant producing ethanol for use in blending with gasoline.

L.B. 571. Passed 1979. Authorizes the governor to enter into agreements with municipalities or counties to build and maintain ethanol plants. The State of Nebraska will have the option to purchase the plant. Creates an Alcohol Plant Fund, to be established with funds transferred from the Highway Trust Fund or as appropriated by the legislature. Increases the state gas tax by one cent to provide additional revenue for the Highway Trust Fund to support the Alcohol Plant Fund.

New Mexico

H.J.M. 15. Passed 1979. Urges the Commodity Credit Corporation of the USDA to consider applications for guaranteed loans from New Mexico companies or organizations to establish pilot projects in New Mexico to convert wood and agricultural waste into alcohol and other energy sources.

H.M. 41. Enacted 1980. Supports the development of alcohol fuel plants and the production and use of Gasohol and Dieselhol; encourages the Energy and Minerals Department to study this energy source.

North Carolina

H.J.R. 1143. Passed 1979. Authorizes the Legislative Research Commission to study the feasibility of producing and distributing alternative fuels to be used for partial substitution of gasoline from agricultural and forest products grown in North Carolina.

S.J.R. 702. Passed 1979. Directs the Energy Division of the Department of Commerce to develop guidelines for the study of producing a partial substitute for gasoline from agricultural and forest products grown in North Carolina.

North Dakota

S.C. Res. 4080. Passed 1979. Urges the federal government to assist and encourage the manufacture and marketing of Gasohol.

Oklahoma

S.B. 428. Enacted 1980. Excludes alcohol produced for use as a motor vehicle fuel from the provisions of the Alcoholic Beverage Control Act; requires federal permits for alcohol production; requires registration of alcohol producers with the Department of Agriculture; requires the department to maintain a list of registered producers.

Oregon

S.B. 927. Passed 1979. Creates energy task forces for solar, wind, geothermal, water, agricultural and forest residue, and alcohol fuels. Also establishes an Alternative Energy Development Commission to prepare comprehensive alternative-resources plans to be submitted to the governor and legislature.

H.B. 2779. Enacted 1979. Allows tax credit on privately financed portion of energy conservation facility based on cost of facility and extent to which facility is being financed by governmental or quasi-governmental body or municipal corporation; credit allowed in each of first two tax years in which credit is claimed shall be 10% of the certified cost of the facility but shall not exceed tax liability of the taxpayer; defines energy conservation facility as any land, structure, or any addition to an existing structure necessarily acquired, erected, constructed, or installed by any person in connection with the conduct of a trade and actually used in the processing or utilization of renewable energy resources; examples of energy resources are wood waste, solar energy, wind power, water power; applies to energy conservation facilities certified on or after 1 January 1987.

H.B. 2780. Enacted 1979. Exempts property used in the production of ethanol from ad valorem taxation if 75% of the fuel produced at the plant is used in making Gasohol; applies to tax years beginning on or after 1 January 1980 but prior to 1 January 1986; Gasohol means a motor vehicle fuel that is a mixture of at least 10% ethanol or methanol not produced from petroleum, natural gas, or coal.

Pennsylvania

S.B. 1011. Signed 1979. Permits production, storage, and use of denatured ethanol for the purpose of providing fuel for personal or business vehicles or machinery by any person upon payment of an annual $25 registration fee; prohibits any sale of the alcohol or utilization by any person other than the producers; requires no posting of bond; requires each licensee to file annual reports with the Pennsylvania Liquor Control Board of monthly production and utilization of the fuel; provides that any violations are subject to the same penalties as other liquor code violations; effective 12 February 1980.

South Carolina

H.R. 3271/S.R. 617. Passed 1980. Encourages the president and Congress to develop the alcohol fuel industry.

South Dakota

H.B. 1008. Signed 1980. Provides for a property-tax assessment credit for residential application of a renewable resource energy system or an ethanol fuel production system; credit is the sum equal to the assessed valuation of the real property with the system minus the assessed valuation without the system; however, the credit shall not be less than the actual installed cost of the system; credit may be applied for three continuous years for residential application and for commercial application of an ethanol fuel production system; at the end of this time, the owner of the real property is entitled to a property-tax assessment credit of 75% of the base credit for the first year subsequent to the termination of the credit period, 50% of base credit for second year, and 25% for third year; adds system producing ethanol for use as fuel to renewable energy resource system entitling owner of real property to a credit; defines "biomass" as an energy source derived from the conversion of organic matter.

H.B. 1205. Signed 1980. Establishes filing requirements for any person refining alcohol for use as an energy fuel, and revises the definition of Gasohol.

Tennessee

H.J.R. 161. Passed 1979. Creates a special joint committee to study the development and use of methanol as an alternative fuel.

H.B. 1778. Signed 1980. Relates to the manufacture, use, and sale of alcohol for fuel purposes.

S.B. 1927. Signed 1980. Relates to the manufacture of alcohol for use as a fuel; provides for a permit and fee.

Texas

H.B. 1803. Enacted 1979. Provides for state loans for establishment of plants to manufacture fuel from renewable resources; $25,000 may be loaned to any one legal entity, and $500,000 may be loaned to a small incorporated business.

H.B. 1986. Enacted 1979. Provides for annual alcohol manufacturing permit of $100. A Texas Legislative Council's preliminary draft provides an alcohol user's license of $10, an alcohol fuel manufacturing license of $25, an agricultural fuel marketing license of $50, and a beverage alcohol manufacturing permit of $1,000.

H.C.R. 230. Passed 1979. Declares that Reps. Bill Keese and Dan Kubiak are the "Co-Fathers of Gasohol in Texas;" recognizes Gasohol as an important new energy resource; declares that the DOE has moved Texas to second place on a priority list of sites for major ethanol plants.

Utah

S.B. 10. Enacted 1980. Allows as a "reasonable allowance" deduction from net income the cost of depletion and depreciation of improvements on facilities used primarily for the production of alcohol fuel.

Virginia

Sub. H.B. 68. Enacted 1980. Authorizes the manufacture of industrial alcohol or ethanol; establishes procedures for the issuance of permits, sales, transportation, and penalties.

H.J.R. 5. Passed 1980. Requests state institutions of higher education to study alcohol fuels; commends the Department of Agriculture and Consumer Services for their efforts to promote alcohol fuel production.

Washington

Sub. H.B. 302. Enacted 1979. Exempts from state excise tax any person who manufactures alcohol to be used in the production of Gasohol for use as motor vehicle fuel; Gasohol is defined as a motor vehicle fuel that contains more than 9.5% alcohol by volume.

H.J.M. 16. Enacted 1980. Urges the president, Congress, secretary of agriculture, and state governor to eliminate regulatory barriers to ethanol fuel

production and to stimulate production with investment
tax credits and rapid amortization schedules.

S.B. 3551. Enacted 1980. Exempts all real and personal
property used primarily for manufacturing alcohol fuel
from property and leasehold taxation for six years;
defines alcohol fuel as any alcohol made from a product
other than petroleum or natural gas, which is used alone
or in combination with gasoline or other petroleum
products for use as a fuel.

Wisconsin

A.B. 777. Signed 1980. Replaces corporate tax
write-off for solar equipment with tax credit; removes
seven-cent-per-gallon excise tax on Gasohol for 18
months, to increase annually thereafter; reduces annual
state fee for producing Gasohol from $750 to $10;
increases bonding authority for the Housing Finance
Authority by $5 million to fund alternative energy
systems for low- and middle-income housing; requires
solar design and lowest total life-cycle cost on all new
construction of state buildings and university residence
halls; directs studies of Gasohol tax exemption, road
and lab tests of Gasohol performance, and feasibility of
producing ethanol from solid wastes; requires
manufacturers of solar equipment to disclose warranties
and to provide notice of eligibility for tax credits to
purchasers.

FEDERAL LEGISLATION: ALCOHOL-FUEL USE[1]

Energy Tax Act of 1978

Status: Passed, Public Law 95-618, Section 221.
Summary: One of five national energy acts, it provides
motor-vehicle fuel excise-tax exemptions on
gasoline-alcohol fuel blends. The exemption is four
cents per gallon of blend, or 40 cents per gallon
($16.80 per barrel) of alcohol used for 10% blends.
Alcohol fuels are also eligible for DOE entitlements
worth approximately $2.10 per barrel of ethanol, or five
cents per gallon, in late 1979.

---

[1]Federal legislation is listed consecutively by public
law number.

Emergency Energy Conservation Act of 1979

Status: Passed, Public Law 96-102.
Summary: Authorizes the president to establish an emergency energy-conservation program; requires the president to submit a gas rationing plan to Congress; and gives the federal government leverage to encourage the states to adopt their own conservation plans.

Surface Transportation Assistance Act of 1978

Status: Passed, Public Law 96-106.
Summary: Provides for the creation of the National Alcohol Fuel Commission (NAFC) to study the ramifications of alcohol fuel use. Includes as an amendment (Section 20) a requirement that the NAFC complete its report within 18 months after coming into existence. Also authorizes an appropriation of $3 million for the NAFC.

Amendment to Surface Transportation Assistance Act

Status: Passed, Public Law 96-106.
Summary: Extends the National Alcohol Fuel Commission's reporting deadline on the potential for alcohol fuels and increases the authorization level of the commission.

Department of Defense Authorization Act of 1980

Status: Passed, Public Law 96-107.
Summary: Requires the U.S. Department of Defense to use ethanol-gasoline belnds in its vehicles to the maximum extent that is feasible and consistent with overall defense needs and sound vehicle management practices.

Transportation Department Appropriations For FY 1980

Status: Passed, Public Law 96-131.
Summary: Appropriates $1.5 million for the National Alcohol Fuels Commission.

Consumer Checking Account Equity Act of 1979

Status: Passed, Public Law 96-221
Summary: Amends the Federal Reserve Act. (Note: Sen. Harry Bellmon's amendment on 26 November 1979, amends Section 236 to extend credit provisions to apply to alcohol fuels; the Equity Act also broadens definition of alcohol to include methanol and ethanol, either for direct use or in a blend with petroleum.)

Gasohol Competition Act of 1980

Status: Passed, Public Law 96-493.
Summary: Amends the Clayton Act to prohibit
restrictions on the use of credit instruments in the
purchase of Gasohol or other synthetic motor vehicle
fuels when there are no similar limitations on
conventional motor vehicle fuels; prohibits unreasonable
discrimination or limitation on the sale, resale, or
transfer of Gasohol or other synthetic fuels.

Comprehensive Oil Pollution Liability and Compensation Act

Status: Passed, Public Law 96-499.
Summary: Subtitle E of Title VIII imposes an import
duty on ethanol imported for use as fuel.

STATE LEGISLATION: ALCOHOL FUEL USE[1]

Alabama

S. 354. Passed 1980. Provides a reduction of three
cents per gallon in the motor-vehicle fuel tax on
Gasohol.

Arkansas

S.B. 454. Enacted 1979. Directs that motor vehicle
fuel in a mixture with at least 10% alcohol is exempt
from the motor-vehicle fuel tax of nine-and-one-half
cents per gallon. For Gasohol to be eligible for this
exemption, the alcohol used in its production must be
manufactured or distilled within Arkansas from
agricultural or forest products. However, Gasohol
produced in other states is also exempt if those states
exempt Arkansas Gasohol.

---

[1]State legislation may be reported as passed, which is
the year it was passed by a vote of the legislature;
signed, which is the year it was signed by the governor;
or enacted, which is the year the legislation went or is
scheduled to go into effect. State legislation is
listed alphabetically by state and consecutively by year
passed, signed, or enacted.

California

A.B. 2882. Passed 1980. Requires the secretary of business and transportation to reimburse the county of San Diego and the city of San Diego from funds appropriated for converting fleet vehicles to straight alcohol fuel use.

S.B. 318. Enacted 1979. Directs the Department of General Services to prepare a plan for utilizing a fuel containing at least 5% alcohol for use in at least 25% of the vehicles maintained by the department.

S.B. 1324. Passed 1980. Exempts Gasohol from sales and use taxes under specific circumstances.

S.B. 1401. Enacted 1979. Recognizes methanol fuel and methanol-gasoline blends as legal fuels in California; directs the Department of Motor Vehicles to establish a 10-year methanol fuel experimental program.

S.B. 1576. Passed 1980. Exempts fuels manufactured from renewable resources--including agricultural crops or commodities--and from organic waste products from state and local sales and use taxes.

S.B. 1626. Passed 1980. Provides that any alcohol produced for use in or as a motor vehicle fuel be taxed as motor vehicle fuel and not be subject to any other taxes; specifies that state requirements for determining whether alcohol is being produced for use as fuel must be the same as the BATF requirements.

Colorado

S.B. 80. Enacted 1978. Creates a nine-member committee to promote the production and use of Gasohol, alcohol, and related industrial hydrocarbons from Colorado agricultural and forest products. The bill appropriates $8,000 for administration.

H.B. 1135. Passed 1979. Directs that motor vehicle fuels that contain at least 10% alcohol that has been derived in Colorado receive a five-cent excise-tax reduction if sold in counties with a population exceeding 200,000. As the availability of Gasohol increases, all state and local vehicles will be required to use Gasohol.

H.B. 1463. Passed 1979. Defines Gasohol as a fuel blend containing at least 10% alcohol manufactured in Colorado from agricultural or forest products. The alcohol must be at least 95% pure. Gasohol is taxed at

two cents per gallon in lieu of the seven-cent-per-gallon motor-vehicle fuel tax. This special rate is effective until 1 July 1985. Creates Gasohol Promotion Fund, to be administered by the Colorado Department of Agriculture's Gasohol Promotion Committee.
H.B. 1607. Signed 1979. Expands the definition of Gasohol to include motor vehicle fuels containing alcohol derived from hydrocarbon or carbon-containing by-products or waste products; grants a reduction in the property tax on facilities for the production of such alcohol; authorizes a five-cent tax exemption to apply to blends of gasoline and alcohol with a purity of at least 95% which have been produced from Colorado products derived from hydrocarbon or carbon-containing by-products waste products or as agricultural or forest products.

Florida

S.B. 903. Signed 1980. Exempts the sale or distribution of certain motor vehicle fuels blended with alcohol from the first seven cents of the gas tax for a designated period; provides for ethanol fuel development tax-incentive credit to be allowed against the tax imposed under the Florida Income Tax Code to corporations that establish a new business or expand an existing business engaged in the distillation of ethanol or manufacture of equipment for the processing and distillation of ethanol for motor vehicle fuels; authorizes the sale at retail service stations of alcohol-blended fuels meeting certain specifications.

Hawaii

S.B. 1906. Passed 1980. Exempts Gasohol from sales tax from 1 July 1980 to 1 July 1985; requires an annual report from the tax director comparing prices and amounts of gasoline and Gasohol sold, together with recommendations for whether or not to continue the exemption incentive.

Idaho

S.B. 1245. Signed 1980. Provides a 50% motor-vehicle fuel tax refund to Gasohol distributors.

Illinois

H.B. 3403. Passed 1980. Directs the Institute of

Natural Resources to implement research, provide information, and hold public seminars to promote increased production and use of alcohol fuels.

S.B. 1518. Passed 1980. Exempts Gasohol from the state occupation and use taxes.

Indiana

S.B. 218. Enacted 1979. Authorizes an exemption from the 4% sales tax of a 10% agriculturally derived ethanol-gasoline blend.

S.B. 315. Signed 1980. Provides for vapor-pressure testing of Gasohol to be conducted in accordance with the methods outlined by the American Society for Testing and Materials.

Iowa

H.F. 491. Enacted 1978. Exempts Gasohol from the state excise tax of 10 cents per gallon; defines Gasohol as containing at least 10% alcohol distilled from agricultural products. The first stage began 1 July 1978, with a total exemption from the motor-vehicle fuel tax of eight-and-one-half cents per gallon; the second stage began 1 July 1979, with an exemption of 10 cents per gallon. This increased exemption (eight-and-one-half cents to 10 cents) was the result of an increase in the motor-vehicle fuel tax by the same rate. The tax exemption on Gasohol ceases 30 June 1983.

H.F. 734. Enacted 1979. Appropriates $50,000 to the Iowa Development Commission for FY 1979 for the promotion of Gasohol and associated by-products; extends the time period for funds appropriated for ethanol research to the end of FY 1980; Iowa State University of Science and Technology is conducting this research project.

H.F. 738. Enacted 1979. Specifies intent of the General Assembly that state agencies shall make every effort to use Gasohol in vehicles operated by the agencies when Gasohol is reasonably available and competitively priced.

S.F. 2376. Signed 1980. Imposes an excise tax of five cents per gallon on Gasohol effective 1 May 1981 and exempts Gasohol from the sales tax when the excise tax is imposed; requires a Gasohol blender's license for persons blending alcohol and motor vehicle fuel on or after 1 July 1983; provides for the payment of tax on stored Gasohol, motor vehicle fuel, and special fuel when an increase in the excise tax occurs, which rate of tax shall be equal to the difference between the old rate of tax and the increased tax rate.

Kansas

H.B. 2324. Enacted 1979. Authorizes a five-cents-per-gallon tax exemption for a 10% 190-proof ethanol-gasoline blend. The ethanol must be produced from grain grown in Kansas. The Act is to be effective 1 July 1979 and the tax exemption shall be reduced one cent per year until no tax exemption remains after 1 July 1984. Requires all motor vehicles owned and operated by the State of Kansas and subdivisions to be operated with a 10% blend of ethanol when reasonably obtainable.

Kentucky

H.B. 838. Signed 1980. Reduces and exempts Gasohol from state taxes.

Louisiana

H.B. 571. Signed 1979. Exempts Gasohol from the state sales and use taxes and motor-vehicle fuel tax; Gasohol is defined as a blend of gasoline containing at least 10% alcohol that has been distilled in Louisiana from agricultural commodities; provides that all state-owned motor-vehicle fuel distributors supply Gasohol for use by state vehicles whenever possible.

S.B. 968. Passed 1980. Creates the Louisiana Gasohol Authority within the Department of Natural Resources and describes its powers, duties, and functions; provides incentives for the use of Gasohol, including exemptions from the state sales and use taxes and the motor-vehicle fuel tax on Gasohol and Gasohol equipment.

Maryland

S.B. 807. Signed 1979. Authorizes a one-cent tax exemption to apply to a blend of 10% ethanol or methanol gasoline.

H.B. 1628. Signed 1979. Requires the secretary of agriculture to study the effectiveness of an ethanol-gasoline mixture; requires the secretary of agriculture to initiate a one-year testing program using Gasohol in eight state-owned vehicles.

H.J.R. 24. Passed 1979. Requests the secretary of agriculture to appoint a commission to study the feasibility of utilizing Gasohol as an alternative fuel.

S.B. 9. Signed 1980. Increases existing motor-vehicle fuel tax reduction of one cent per gallon for Gasohol to four cents per gallon until 1 May 1981, when it reverts to one cent per gallon less than other motor vehicle fuels; effective 22 April 1980.

S.B. 492. Signed 1980. Includes Gasohol in the Gasoline Products Marketing Act, which regulates retail and wholesale trade practices and participants; prohibits distributors from preventing the purchase of Gasohol by credit card; amends act title to include Gasohol.

S.B. 675. Signed 1980. Exempts ethanol or methanol sold as motor vehicle fuels but not mixed with gasoline from motor-vehicle fuel tax and retail sales tax; provides that these exemptions automatically expire 30 June 1982; effective 1 July 1980.

Michigan

S.B. 480. Passed 1980. Imposes a tax of six cents per gallon on Gasohol.

Minnesota

H.F. 8. Passed 1980. Reduces tax on gasoline by two cents per gallon if it is 10% agriculturally derived ethanol that is at least 165 proof; appropriates 5% of revenue from gasoline taxes collected to the agricultural alcohol-fuel tax fund for testing, evaluating, and promoting alcohol fuel use.

Mississippi

H.B. 366. Signed 1980. Allows the blending of gasoline and alcohol at various facilities in the state which will be designated and licensed by the motor vehicle comptroller.

Missouri

H.B. 72. Passed 1979. Authorizes the Department of Natural Resources to analyze the potential for increased utilization of coal, nuclear, and solar energy, resource recovery and reuse, energy-efficient technologies, and other energy alternatives, and to make recommendations for the expanded use of alternative energy sources and technologies.

Montana

H.B. 402. Passed 1979. Reduces the tax exemption for Gasohol by two cents for each of three succeeding two-year periods; the remaining one-cent tax exemption expires in 1989.

Nebraska

L.B. 776. Passed 1971. Establishes the Agricultural Products Utilization Committee to promote Gasohol research and development, and to analyze Gasohol marketing and testing. Creates the Grain Alcohol Fuel Tax Fund with an initial appropriation of $40,000 and a provision whereby one-eighth of the motor-vehicle fuels tax, which is refundable to nonhighway users, must go to promote the committee's activities. Also provides for a three-cent tax credit for Gasohol sales and provides for a tax-credit review after 10 million gallons of Gasohol have been sold.

L.B. 1208. Passed 1972. Makes changes in L.B. 776. States that to qualify as Gasohol, the ethanol must be at least 190 proof. Also directs the committee to sponsor research and development of industrial uses of by-products resulting from the amended L.B. 776; increases the tax credit from three cents to five cents on Gasohol, and increases the legislative tax-credit review limitation from 10 million to 20 million gallons of Gasohol sold.

L.B. 74. Enacted 1979. Requires that the Department of Roads implement a program using Gasohol in its vehicles to the extent that Gasohol supplies are available. Gasohol must contain Nebraska-produced alcohol.

L.B. 876. Passed 1980. Increases the amount of funds transferred to the Agricultural Alcohol Fuel Tax Fund.

New Hampshire

H.B. 201. Signed 1979. Authorizes a Gasohol tax
exemption of five cents a gallon to apply to a 10% blend
of alcohol manufactured in New Hampshire that has a
purity of 99% and has been derived from agricultural
commodities and forest products.

New Jersey

S.R. 22. Passed 1979. Directs that a joint feasibility
study by the director of the Division of Energy Planning
and Conservation and the director of the Division of
Purchase and Property be undertaken regarding
use of Gasohol on a major scale, and that if the
determination is favorable, to undertake a pilot program
and utilize Gasohol in state vehicles.

A.R. 3034. Passed 1979. Directs the Energy and Natural
Resources Committee of the General Assembly to study
use of Gasohol and other alcohol-based fuels on a major
scale.

New Mexico

S.J.M. 9. Passed 1978. Requests the Energy and
Minerals Department to study the feasibility of using
Gasohol in New Mexico.

H.M. 41. Enacted 1980. Supports the development of
alcohol fuel plants and the production and use of
Gasohol and diesel-alcohol blends; encourages the Energy
and Minerals Department to study this energy source.

S.B. 39. Signed 1980. Provides a temporary gross
receipts and compensating tax deduction and a gasoline
tax deduction as incentives for the production and sale
of Gasohol.

New York

S.B. 9860-A. Passed 1978. Directs the commissioner of
general services to conduct a study of the feasibility
of using Gasohol for state-operated vehicles through a
comprehensive road test.

S.B. 2393. Passed 1979. Directs the commissioner of
general services to conduct an experimental program to

test the feasibility of using a mixture of gasoline and alcohol as fuel for state-operated motor vehicles.

North Carolina

H.J.R. 1143. Passed 1979. Authorizes the Legislative Research Commission to study the feasibility of producing and distributing alternative fuels from agricultural and forest products grown in North Carolina, to be used as a partial substitute for gasoline in North Carolina.

North Dakota

S.C.R. 4035. Enacted 1975. Urges various state facilities to conduct research to determine the feasibility of using grain alcohol as a renewable alternative energy source.

H.B. 1384. Signed 1979. Establishes an Agriculture Products Utilization Committee funded by an eight-cent-per-gallon gasoline-tax reduction, with $200,000 appropriated from 1 July 1979 to 30 June 1980.

S.B. 2338. Signed 1979. Lowers the fuel tax for Gasohol to four cents from eight cents; exemption applies to a blend of 10% agriculturally derived ethanol with a purity of 90% and unleaded gasoline.

S.C.R. 4080. Passed 1979. Urges the federal government to assist and encourage the manufacture and marketing of Gasohol.

Oklahoma

S.B. 248. Enacted 1979. Exempts from regular motor-vehicle fuel tax blends of motor vehicle fuel and ethanol containing at least 10% by volume of 199-proof ethanol distilled from agricultural products; Gasohol is taxed at 0.08 cents per gallon rather than 6.58 cents per gallon as other motor vehicle fuels are.

Oregon

H.B. 2770. Passed 1979. Requires use of Gasohol in certain state-owned vehicles to the maximum extent commercially feasible; effective January 1980.

Rhode Island

H.R. 7891. Passed 1978. Requests that the director of transportation conduct tests to determine the feasibility of using a gasoline-alcohol fuel blend.

H.R. 5998. Passed 1979. Requests that the director of transportation conduct experiments with state-operated motor vehicles to determine the feasibility of using Gasohol in those vehicles.

South Carolina

H.B. 2443. Enacted 1979. Provides that Gasohol be sold tax-free until 1 October 1979; imposes a six-cent-per-gallon tax from 1 October 1979 to 1 July 1985, and a seven-cent-per-gallon tax from 1 July 1985 to 1 July 1987; provides for removal of these incentives if loss of revenue totals $5 million.

South Dakota

H.B. 1064. Enacted 1979. Reduces the excise tax on Gasohol; exempts ethanol that is to be blended with motor vehicle fuel to produce Gasohol from the sales and use taxes; provides a tax credit on the motor vehicle fuel blended with ethanol to produce Gasohol; Gasohol is defined as motor vehicle fuel containing a minimum of 10% of ethanol derived from agricultural or forest products blended with gasoline.

S.B. 175. Enacted 1980. Defines Gasohol as a motor vehicle fuel containing a minimum of 10%, at least 198-proof ethanol by volume blended with gasoline. The ethanol must be derived from agricultural or forest products.

Tennessee

H.J.R. 161. Passed 1979. Creates a special joint committee to study the development and use of methanol as an alternative fuel.

H.B. 1778. Passed 1980. Establishes regulations for the manufacture, use, and sale of alcohol for fuel purposes.

S.B. 248. Enacted 1979. Exempts from regular motor-vehicle fuel tax blends of motor vehicle fuel and

ethanol containing at least 10% by volume of 199-proof
ethanol distilled from agricultural products; Gasohol is
taxed at 0.08 cents per gallon rather than 6.58 cents
per gallon as other motor vehicle fuels are.

Texas

H.C.R. 230. Passed 1979. Declares that Reps. Bill
Keese and Dan Kubiak are the "Co-Fathers of Gasohol in
Texas;" recognizes Gasohol as an important new energy
resource; declares that DOE has moved Texas to second
place on a priority list of sites for major ethanol
plants.

H.B. 1986. Enacted 1979. Provides for annual
alcohol manufacturing permit of $100. A Texas
Legislative Council's preliminary draft provides an
alcohol users license of $10, an alcohol fuel
manufacturing license of $25, an agricultural-fuel
marketing license of $50, and a beverage alcohol
manufacturing permit of $1,000.

Utah

S.B. 11. Enacted 1980. Provides a reduction of four
cents per gallon in the state motor-vehicle fuel tax on
Gasohol beginning 1 July 1980.

Washington

H.B. 1568. Enacted 1980. Authorizes the state finance
director to set policies that would, among other things,
require widespread use of Gasohol and other cost-
effective alternative fuels in all state-owned motor
vehicles.

S.B. 3629. Enacted 1980. Exempts alcohol of any proof
that is sold for use solely as fuel in vehicles from the
motor-vehicle fuel and special-fuel taxes through 31
December 1986.

Wisconsin

S.B. 79. Enacted 1979. Provides that a holder of a
limited manufacturer's permit may use or sell
intoxicating liquor only if it is rendered unfit for use
as a beverage and is used or sold for use in an internal
combustion engine.

A.B. 456. Passed 1980. Subjects Gasohol to the same type of regulation as gasoline; includes project standards, inspection, and other legislative activities of the Department of Industry, Labor and Safety Regulations.

A.B. 777. Passed 1980. Replaces corporate tax write-off for solar equipment with tax credit; removes seven-cent-per-gallon excise tax on Gasohol for 18 months, to increase annually thereafter; reduces annual state fee for producing ethanol fuel from $750 to $10; increases bonding authority for the Housing Finance Authority by $5 million to fund alternative energy systems for low- and middle-income housing; requires solar design and lowest total life-cycle cost on all new construction of state buildings and university residence halls; directs studies of Gasohol tax exemption, road and lab tests of Gasohol performance, and feasibility of producing ethanol from solid wastes; requires manufacturers of solar equipment to disclose warranties and to provide notice of eligibility for tax credits to purchasers.

Wyoming

H.B. 114. Enacted 1979. Authorizes a motor-vehicle fuel tax decrease from eight cents to four cents for Gasohol until 1 July 1984.

REFERENCES

Federal and state legislation as listed in the text.

Merritt, Richard. 1981. Renewable Fuels Association, Washington, D.C. Personal communication.

Moorer, Richard. 1981. U.S. Department of Energy, Office of Alcohol Fuels, Washington, D.C. Personal communication.

8
International Alcohol Fuel Production

INTRODUCTION

Alcohols provide an alternative to petroleum products both for liquid fuels and for feedstocks in the chemical industry. Alcohol fuels can be produced by established technology. Feedstocks for alcohol fuels are agricultural, forest, and waste resources. Although there is concern that crops grown for alcohol production will have an adverse impact on food production, this need not be the case. Careful and coordinated planning is essential and should be integrated into national goals aimed at developing self-reliance regarding food and energy resources. Alcohol fuels have the potential to provide not only the developed nations, but also those in various stages of development, an opportunity to produce a strategically important energy need.

Potential uses of alcohol fuels include (1) engine fuels, either straight (neat) or as blends; (2) feedstocks for the chemical industry; (3) feedstocks for single-cell protein; (4) fuel for fuel cells; (5) fuel for lighting and cooking in rural nonelectrified areas.

While countries on all continents are considering alcohol fuels, only one country, Brazil, has made a full-scale commitment to developing such an industry to date. In the mid-1970s, Brazil imported 80% of its crude oil. However, as a result of rising oil prices, the impact on their balance of trade, and the potential shortage of supplies, the government developed a National Alcohol Program in 1975. Before 1975, about 600,000 m^3/yr of ethanol were produced; in 1978-1979, this was increased to 2.6 million m^3/yr, and production for 1979-1980 was expected to be 4 million m^3/yr (Brazil Ministry of Industry and Commerce n.d.).

The social, economic, and political implications of the Brazilian program include:

1. providing additional jobs in the agricultural sector;
2. helping slow urban migration;

3. improving income distribution;
4. balancing development geographically;
5. providing a more favorable balance of trade;
6. supporting development of domestic production facilities; and
7. providing independence from foreign sources of energy and chemical feedstocks.

Problem areas in the program are being investigated by various Brazilian agencies and institutions. These areas include:

1. reduction of investment and operating costs;
2. coordination of site selection of agricultural and industrial operations;
3. improvement of the agricultural sector's ability to provide feedstocks;
4. improvement of the fermentation distillation process, particularly in the area of energy use;
5. more extensive development of the alcohol distribution program, emphasizing industrial decentralization;
6. improvement of the economic viability of independent distilleries; and
7. examination of market potential of ethanol as a chemical feedstock.

Despite these issues, Brazil has made rapid progress toward energy independence by using alcohol fuels to replace imported crude oil. Brazil's activities will be watched with great interest by countries with sufficient agricultural resources to grow not only food, but feedstocks for fuel production.

Several nations, in addition to the United States, have begun developing a limited alcohol fuels industry. Activities range from research and planning to operating plants of varying capacities that are meeting limited local, regional, and national needs for liquid fuels. In addition to the following summaries, there are unconfirmed reports of alcohol fuel production and use in Cuba and the Soviet Union.

AFRICA

Egypt

Blackstrap molasses from sugarcane processing is the feedstock used by the Egyptian Sugar and Distillation Co. for ethanol production. Analysis of the molasses shows that it comprises principally sucrose, glucose, and fructose, with traces of arabinose and melezitose. There appears to be a higher than usual percentage of aconitic acid in Egyptian molasses. It ranges from 5.9% - 6.4%. Egyptian ethanol production involves:

1. clarifying the diluted molasses in centrifugal clarifiers;
2. adding yeast and fermenting the mixture with the aid of partial aeration and added nutrients;
3. running the mash into yeast separators;
4. washing and drum-drying the separated yeast to produce fodder yeast with 40% protein; and
6. distilling the mash.

The nutrients used are ammonium phosphate, superphosphate, and urea. The fermentation cycle is 14 hours. Thermal and chemical clarification of the molasses helped solve problems of yeast flocculation. The problem of cooling large fermentors in a tropical climate such as Egypt's was solved by using internal coils, other surface-extending media, and external heat-plate exchangers.

About 45 million L/yr of ethanol are produced by the Egyptian Sugar and Distillation Co. The 30 million L/yr of 195-proof ethanol produced is used in the following ways:

1. for production of vinegar, using surface and submerged fermentation processes (80% of this vinegar is used in glacial acetic-acid production);
2. for production of eau de cologne, perfume, and cosmetics;
3. for production of alcoholic beverages;
4. for production of ether;
5. for production of ethyl-acetate solvent in an esterification plant; and
6. for limited use in the pharmaceutical industry and hospital and chemical laboratories.

About 10 million - 15 million L/yr are available for export. About 3 million L/yr of specially denatured ethanol from 190 proof - 195 proof are used as a chemical feedstock and solvent. About 14 million L/yr of completely denatured 190-proof alcohol are used for heating and lighting. Egypt has not used ethanol as a motor vehicle fuel--either alone or blended with gasoline--although this is being explored (Madi 1979).

Kenya

Kenya has one ethanol production plant at Kisumu scheduled to come on-stream in 1982 with a capacity of 20 million L/yr (approximately 5.28 million gal/yr). Fuel oil will provide process heat and the feedstock will be molasses. Construction costs are estimated at K Sh1 billion--approximately US$138 million--compared to a World Bank estimate for ethanol plant construction costs of US$100/liter capactiy/day; which for a 20 million L/yr plant is approximately US$5.5 million (Kohli 1980). The World Bank estimate, which is the equivalent of 95¢/gal capacity, is for early 1980. In early 1981, construction costs in the U.S. were approximately US$2/gal capacity (Curet 1981). Two additional plants are being planned, one in Muhroni scheduled to come on-stream in 1983 with a capacity of 18 million L/yr. The other, in South Nyanza, is scheduled to come on-stream in 1985 with a capacity of 45 million L/yr. These two plants are expected to use molasses as a feedstock and burn bagasse for process heat. The ethanol will be mixed with gasoline for use as a motor vehicle fuel (Autonews 1981).

South Africa

African Chemicals of Johannesburg has been producing ethanol from maize for use as an industrial chemical for several years. There is limited production of ethanol from sugarcane for use as a fuel extender (Hyde 1980).

The government is exploring the possibility of converting the Makatini Flats in northern Zululand into cassava plantations, using large-scale plant nurseries and a seedling production process involving cloning. The cassava would be used as a feedstock for ethanol production in 13 plants scheduled to be built in the area that would produce 137 million gal/yr (Brown 1980).

Sudan

The Sudanese government is considering a project to utilize the molasses residue from established and planned sugarcane refineries. Government forecasts expect sugarcane refining to reach 870,000 metric tons/yr in the 1980s, with 305,000 metric tons of molasses available for export and/or alcohol fuel production. A government study suggested a 16,000 metric tons/yr alcohol plant be built at West Sennar next to the sugar refinery, with a larger plant to be built at Kenanar (Brown 1979).

Tanzania

In 1977, Vogelbusch Gesselschaft of Vienna, Austria, prepared an economic study commissioned by the Tanzanian government on molasses utilization. One of the conclusions was that production of alcohol fuel may be viable. The Tanzanian market for alcohol fuels was estimated at between 4.25 million L/yr and 7.75 million L/yr. It was also recommended that some of the proposed ethanol production be used as a feedstock for polymer

products, such as water pipe. The Sugar Development
Corp. in collaboration with the Tanzania Petroleum
Development Corp. is planning an alcohol fuel project at
Kagero. When the molasses export commitment ends in
1985, additional plants are planned for Kilombero and
Mtibwa (Mungai 1979).

ASIA/SOUTH PACIFIC

Australia
     The first Australian alcohol fuels pilot plant,
developed by Biotechnology Australia Proprietary, went
on stream in 1980, and is being expanded to produce
2.5 million L/yr. Expansion and future plants are being
managed by a consortium called Australian Ethanol Fuels
Proprietary, which includes Biotechnology Australia
Proprietary, backed by Ampol Petroleum, Australian
Mutual Provident Society, and Thomas Nationwide
Transport. The expansion plans for the plant, which is
located near Sydney, call for a $A1 million budget
through 1981. Wheat is being used as the feedstock, and
the production is accomplished with a
continuous-fermentation process developed by David
MacLennon of Biotechnology Australia Proprietary. The
process is adaptable for sugar and starch crops,
including fodder beets, sugar beets, and cassava.
Future plants will utilize the most readily available
feedstock. Australia hopes to provide 15% of its liquid
fuel needs from alcohol fuels by 1985 (Business
Publishers 1979a).
     Scientists at the Australian Commonwealth
Scientific and Industrial Research Organization have
developed a solid-phase fermentation process for
producing ethanol from sugar beet feedstocks that
substantially reduces production and capital costs. The
process involves pulping the beets and pressing them
through rollers, which releases 65% of the sugar. The
organization estimates that a plant using this process
would require half the capital costs of a fermentation
plant using conventional processes.
     Researchers at the Appropriate Technology and
Community Environment Research Center have experimented
with a chemical emulsifier that allows ethanol to be
mixed with diesel fuel and used in standard diesel
engines. Tests using mixtures of up to 30% ethanol and
70% diesel fuel have demonstrated unaltered performance.
     The Shell Co. has plans to produce Petranol, a
blend of 10% ethanol and 90% supergrade petrol.
Sugarcane and molasses will be used as feedstocks for
ethanol production. Petranol will be test marketed
through Shell service stations in the Mackay region of

Queensland. The product is being subsidized and will be sold at prices that will be competitive with petrol.

Australian Cassava Products, a joint venture, will construct and supervise a cassava plantation near Maryborough in Queensland and has $A15 million allocated for development. Initial plans are to supply starch to the food and chemical industries, with development of an ethanol production facility by 1986 at an estimated cost of $A10 million - $A15 million (Commonwealth of Australia 1980).

Fiji

The economy of Fiji is based on sugar. Sugar and molasses account for 65% of the country's foreign exchange income. Two studies on the economic feasibility of alcohol fuel production have been prepared by the Fiji Sugar Corp., which is 98% government-owned, and operates the sugar refineries. The conclusions were that--as of 1979--use of molasses for alcohol fuel production was not economically feasible. Potential benefits included improving the balance of trade and creating job opportunities in rural areas. The economics are being monitored, and Fiji expects to move into alcohol fuel production in the future (Karan 1979).

India

India began fermentation ethanol production in 1931, with about 20 plants with a combined production capacity of approximately 3.7 million L/yr. Molasses has traditionally been used as the feedstock. In 1978 about 100 plants had a total production capacity of about 600 million L/yr, although actual production output was closer to 500 million L/yr. Three hundred and ten million liters were used as a chemical feedstock in industry, 170 million liters were used as potable alcohol, and 20 million liters were exported.

The following production process is used:
1. the molasses is diluted and acidified with sulphuric acid in mild-steel fermentation tanks;
2. yeast and a nutrient, such as ammonium sulfate, are added;
3. fermentation proceeds and is completed in about 36 hours (the fermentation vats are cooled as necessary by spraying their outside walls with cold water); and
4. distillation takes place in two stages, first in an analyzer column, then in a rectifier column producing 95% ethanol.

If anhydrous ethanol is desired, the azeotrope is broken with benzene, and distillation is continued. Problems facing India's ethanol industry include:

1. seasonal fluctuation in the availability of molasses;
2. a corresponding fluctuation in the availability of ethanol for the chemical feedstock industry, which requires a continuous supply to maintain efficient production processes;
3. ethanol recovery;
4. inadequate pollution control measures; and
5. a cost for molasses that is too high in relation to the government-controlled price for ethanol.

The primary use of ethanol in India has been as a chemical feedstock rather than a vehicle fuel. Ethanol is being considered as a feedstock for the production of acetaldehyde, acetic acid, acetone, butanol, and 2-ethyl hexanol (Sharma 1979).

Japan

Japan's ethanol production has focused on industrial uses of ethanol--for example, in the cosmetics industry or as a chemical solvent. Molasses is the primary feedstock, but sweet potatoes, potatoes, corn, and sugar beets are also used. The National Alcohol Plant at Chiba in Tokyo, for instance, uses molasses as a feedstock, utilizes a batch production process, and recovers a small amount of carbon dioxide (CO_2) as a by-product. Of the 64.8 million liters of fermentation alcohol used in FY 1978 (April 1978-May 1979) 13.8 million liters were used in the chemical industry; 39.7 million liters were used in the food industry; and 11.3 million liters were used in other industries, such as the medical industry (Yokoi 1979).

A special council is being established in the Ministry of International Trade and Industry to plan the development of alcohol fuels in Japan. Expected feedstocks will include cassava, seaweed, and garbage. The program is being developed in cooperation with Brazil, Indonesia, the Philippines, and Thailand. The Japanese government expects to be using E10 as a vehicle fuel by 1985 (Nagata 1980).

New Zealand

In 1979, 85% of New Zealand's oil was imported; the transportation sector used 70% of the country's total oil supplies. Because of the international oil situation, the government has established a target of more than 50% self-sufficiency in transport fuels by 1987. The Liquid Fuels Trust Board was established to develop a program to achieve this goal.

The major thrust of the program is development of compressed natural gas and liquid petroleum gas, but methanol is also emphasized to some extent. The board has authorized several studies, which will deal with areas such as: (1) methanol-blend specifications and distribution; (2) methanol toxicity; (3) methanol corrosion; (4) single-vehicle and fleet trials using a 15% methanol-85% gasoline blend (M15); (5) methanol use in diesel engines; and (6) assessment of modifications to existing gasoline-engineered vehicles (Liquid Fuels Trust Board 1979).

A report, "The Potential of Energy Farming for Transport Fuels in New Zealand," has been completed. It examined and recommended energy farming using both sugar beets and fodder beets for ethanol production and the radiata pine for methanol production to provide New Zealand with domestically produced liquid fuels for use by the transportation sector. (Ad Hoc Committee on National Development 1980; New Zealand Energy Research and Development Committee 1980).

Papua New Guinea

Construction began in early 1980 on a 2 million L/yr ethanol plant in the isolated Mount Hagen area. Davy Pacific of Australia is building the plant. It will use cassava as the feedstock and a pyrolytic process to produce ethanol. The fuel will be for local use. Dr. Ken Newcomb, energy planner for Papua New Guinea, Department of Minerals and Energy, in Konedobu, Papua New Guinea, is supervising the project (Tatum 1980).

Philippines, The

The Philippine government has made a commitment to expand the country's fermentation alcohol industry to include production of "Alco-gas." In 1981 the government expects to produce 5.8 million gallons of ethanol; by 1989 they expect to replace 24.4 million gal/yr of crude oil with fermentation ethanol. Sugarcane, the primary export commodity, is expected to be the feedstock. Molasses, cassava, and sweet potatoes are also being considered as potential feedstocks. The government has established a National Alcohol Commission and has entered into an agreement with the Brazilian government for technical assistance. Current production uses molasses as a feedstock, and a batch fermentation process. Some ethanol plants purchase molasses from the sugar mills; others are integrated into the sugar mill operation. One such plant uses steam provided by the sugar mill to burn the bagasse waste from the sugarcane for heat. Animal feed coproducts are frequently

227

recovered from the stillage and sold (Guerrero 1980;
Brown 1980).

Thailand
The government of Thailand is examining bids for
construction of alcohol fuel plants. One specification
is that the plant must be able to use various
feedstocks. Therefore, whichever crop is in surplus at
a particular time can be used as a feedstock. Potential
feedstocks include cassava, maize, rice, sugarcane, and
molasses. According to the government, one reason for
examination of establishment of an alcohol fuel industry
is to stabilize agricultural commodity prices (Brown
1980).

EUROPE

Austria
Austria has a grain surplus of 200,000 - 300,000
tons/yr, which is exported to Eastern Europe. The Fuel
from Biomass Project, sponsored by the Austrian Ministry
of Science and Research, is studying the feasibility
of constructing a demonstration alcohol fuel plant using
this surplus grain as a feedstock. The stillage
coproduct would be used for livestock feed, which would
reduce the demand for imported soybean and fishmeal
(Brown 1980).

Sweden
The Volvo Co. and the Swedish government started a
test program in 1978 to ascertain the effectivenes of
pure methanol as a motor vehicle fuel. Special engines
have been built, and methanol is being produced from
wood wastes. At the end of the test (which may continue
through 1985), a decision will be made as to whether to
convert to methanol as the primary motor vehicle fuel
and to retool automotive factories to produce engines
that can utilize methanol (Darvelid 1980).
Several studies are examining ethanol production.
These studies have been designed to ascertain the
feasibility of producing ethanol from sugar beets as
well as other selected feedstocks and of using the
Karpalund sugar factory for ethanol production. The
Swedish Methanol Development Co. (SMAB), in conjunction
with the Swedish Sugar Co., is studying ethanol
production processes. SMAB is also collaborating with
Alfa-Lava; the National Swedish Farmers' Association;
the Skaraborg County potato growing association; the
Skaraborg County council; the Toreboda local council;
and the Skaraborg County development fund to establish a
pilot ethanol production plant using a process developed

by Alfa-Lava. Feedstocks would be surplus potatoes and grains (Swedish Commission for Oil Subsitution 1980).

West Germany

The West German government and the Volkswagen Co. are cooperating on a 3 million-kilometer vehicle test using a blend of 15% methanol-85% low-lead gasoline (M15).

The cars are being tested under different conditions, and in different altitudes and locations, including the Alps and the Arctic. Initial results indicate that starting, cold-weather performance, fuel economy, exhaust emission, service life of vehicle components, safety, and driveability are satisfactory (Citizen's Energy Project n.d.).

SOUTH AND CENTRAL AMERICA

Brazil

In November 1975, after paying US$3.5 billion for oil imports, Brazil announced the beginning of PROALCOOL--Programa Nacional do Alcool (National Alcohol Program). The Cabinet-level National Energy Commission was established, and the government approved a US$5 billion appropriation for the program through 1985. Project goals set in 1975 to be achieved by 1980 were 3.6 billion L/yr. That goal was surpassed with production of 4.1 billion L/yr.

The second phase of the program (1980-1985), has a production goal of 10.7 billion L/yr of alcohol. This would include:

1.   3.1 billion L/yr of anhydrous alcohol for blending with gasoline;
2.   6.1 billion L/yr of anhydrous alcohol for direct use as a fuel; and
3.   1.5 billion L/yr for use as a chemical feedstock.

Additional production needed to meet the second phase production goals will be produced by independent plants not associated with sugar mills. This phase is expected to require a US$5 billion investment. It will create 350,000 new jobs, and 1.7 million hectare of new agricultural land will be developed.

The 1985-1990 goal is 105 billion L/yr of alcohol--about two-thirds of the nation's gasoline consumption.

Ethanol costs less than gasoline in Brazil. Inexpensive and plentiful labor, high gasoline taxes, and government subsidies for the National Alcohol Program are major influences on the economics of ethanol production and use.

As of mid-1981, however, the Brazilian government has trimmed slightly its subsidy of the alcohol fuel industry for the following reasons: (1) to promote more private sector involvement; (2) as a reaction to recession-type international economic pressures, which affect national economics; and (3) to bring alcohol fuel and gasoline closer in price to emphasize to the people the need to conserve on all forms of energy, including alcohol (Lima 1981).

A direct analogy cannot be made between the situation in Brazil and that in the United States. The Brazilian government has chosen to make a high initial investment in an industry that they expect to considerably decrease their dependence on imported crude oil, while the United States is examining various alternatives and has a more limited approach to industry subsidies.

The primary feedstocks have been molasses and sugarcane, partly because the alcohol fuel industry was originally based on the country's surplus sugarcane. Cassava, babassu(coconut), sorghum, and sweet potatoes are being seriously considered or used in production to some extent. Cassava has a higher yield of alcohol per ton than sugarcane--160 - 185 L of alcohol/processed ton of cassava and 60 - 70 L of alcohol/processed ton of sugarcane. However, sugarcane has a better yield of liters of alcohol per hectare than cassava.

While Brazil is using plants with differing production capabilities, the emphasis has been on small-scale plant development to boost local economies in appropriate regions. Initial production was handled by expansion of existing sugarcane manufacturing plants, with a limited number of new plants. The Cuevelo Distillery, for example, produces 60 m^3/day of anhydrous ethanol, using manioc as the feedstock. The manioc was planted on previously uncultivated land. This project has generated about 2,000 new jobs in a previously depressed area. New distilleries' production ranges from 30m^3 - 650 m^3/day, with the average about 120 m^3/day.

Existing plants generally use a batch fermentation process. New plants will press the sugar juice out of the cane and run it into continuous-fermentation vats. Centrifugal separation systems recover about 90% of the yeast. The fermented solution is processed through a series of distillation columns fueled by the bagasse.

The emphasis has been on producing 199-proof anhydrous ethanol, which tests indicate blends best with gasoline. Most Brazilian engines operate with a fuel-air ratio above stoichiometric, which allows engines to perform well with ethanol-gasoline mixtures of up to 20% ethanol. Ethanol also has raised the

octane rating from about 70.3 Motor Octane Number (MON)
to 80.3 MON. Work is being done to convert vehicles
so that ethanol that is less than 196.4 proof can be
used; this ethanol is much more economical to produce.
The first such vehicles came off the assembly line in
late 1979 (Business Publishers 1979a). The
manufacturing rates for cars that can run on straight
ethanol are:

1. for 1980, 250,000 new cars and 80,000
   converted cars;
2. for 1981, 300,000 new cars and 90,000
   converted cars; and
3. for 1982, 350,000 new cars and 100,000
   converted cars (Lima 1980).

From its inception, Brazil's program has had a
strong research and development component. Initially,
the emphasis was on road tests. These showed acceptable
results, but more recently the major focus has been on
(1) cellulose residue fermentation using acid
hydrolysis; (2) conversion of the waste from the
fermentation process into usable products; and (3)
alcohol fuel use in diesel engines. Commercialization
options being considered for alcohol fuel replacement of
diesel fuel include:

1. use of hydrated alcohol with an additive to
   replace dieselfuel;
2. blending 4% alcohol with diesel fuel;
3. use of a dual injection system with 40%
   alcohol and 60% diesel fuel; and
4. replacement of diesel engines with explosion
   engines fueled by hydrated alcohol (Lima
   1980).

Costa Rica

In 1979, the Central American Research Institute
for Industry prepared a feasibility study on production
of ethanol in five Central American countries: Costa
Rica, El Salvador, Guatemala, Honduras, and Nicaragua.
The conclusions included the following:

1. blends of anhydrous ethanol and gasoline are
   technically feasible up to a ratio of 1:3 in
   conventional automobiles, although a ratio of
   1:5.6 was used in the study;
2. blending and handling do not require special
   procedures;
3. ethanol improves the octane number of the
   mixture;
4. ethanol-gasoline mixtures do not cause
   corrosion in engines;
5. production of sugarcane would have to be
   increased 20% - 25%, which is agriculturally

   feasible in all the countries except possibly
   El Salvador;
6. cassava could be used as an alternative
   feedstock, in off-sugarcane seasons;
7. ethanol can be used as a blend with gasoline
   for motor vehicle fuel and as a chemical
   feedstock--polyvinyl chloride monomer was
   recommended; and
8. fluctuations in the world price of sugar could
   have an adverse impact on the economic
   feasibility of ethanol production from
   molasses feedstocks (Ingram 1979).

El Salvador
  See Costa Rica.

Guatemala
  See Costa Rica.

Honduras
  See Costa Rica.

Jamaica
  The Jamaican Sugar Industry Research Institute, the
Organization of American States, and Tegri-Tecnica
Agro-Industrial, Ltd. of Brazil conducted a feasibility
study of production of alcohol for blending with
gasoline for motor vehicle fuel.
  Jamaica's rum distilleries operate on a purchase
order basis, and are not fully utilized. The
feasibility study noted that the off-production periods
for rum could be used for alcohol fuel production.
However, the study concluded it would not be
economically feasible to use molasses as a feedstock,
because of its uses in rum for export. If crude oil
prices continue to increase and the high capital costs
can be lowered, it may be economically viable to produce
alcohol fuel from sugarcane (Sangster 1979).

Nicaragua
  See Costa Rica.

Paraguay
  A 22 million L/yr ethanol plant that will use
sugarcane as a feedstock is being constructed in
Troche. The government-owned facility is being built at
a cost of US$18.5 million, and is expected to be in
full-scale production by 1984. The ethanol will be
distributed throughout the country as a gasoline
extender. Alcohol fuel is a government-operated
monopoly in Paraguay (Caniza 1980; Knapps 1981).

REFERENCES

Ad Hoc Committee on National Development. 1980. Growth
Opportunities in New Zealand. Wellington, New Zealand:
Ad Hoc Committee on National Development.

Autonews. 1981. "The Great Gasohol Goof." Pp. 12-13.
Nairobi, Kenya.

Brazil Ministry of Industry and Commerce. n.d.
Alcohol. [Available from Fundacao de Tecnologia
Industrial, Av. Venezuela, 82 70 andar, 20081 Rio de
Janeiro, RJ, Brazil.]

Brown, Lester R. 1980. "Food or Fuel; New Competition
for the World's Cropland." Washington, D.C.:
Worldwatch Institute.

Brown, O.M.R. 1979. "Power Alcohol in the Sudan - A
Case Study." Presented at Workshop on Fermentation
Alcohol for Use as Fuel and Chemical Feedstock in
Developing Countries; 26-30 March 1979; Vienna,
Austria. Sponsored by the United Nations Industrial
Development Organization.

Business Publishers. 1979a. Solar Energy Intelligence
Report. 16 April. Silver Spring, MD: Business
Publishers.

Business Publishers. 1979b. Solar Energy Intelligence
Report. 15 October. Silver Spring, MD: Business
Publishers.

Caniza, E. 1980. Embassy of Paraguay, Wasington, D.C.
Personal communication.

Citizens' Energy Project. n.d. Gasohol. Washington,
D.C.: Citizens' Energy Project. Report series no. 24.

Commonwealth of Australia. 1980. "Australian Energy
Reports." Canberra, Australia: Commonwealth of
Australia.

Curet, Donald. 1981. Ethanol Production Consultant,
Idaho Falls, ID. Personal communication.

Darvelid, D. 1980. Agriculture Office, Embassy of Sweden, Washington, D.C. Personal communication.

Guarrero, Perfecto. 1980. Scientific Attache, Embassy of The Republic of the Philippines, Washington, D.C. Personal communication.

Hyde, J. 1980. Science and Technology Office, Embassy of the Republic of South Africa, Washington, D.C. Personal communication.

Ingram, Ludwig. 1979. "Fuel and Chemical Feedstock from Sugar Cane in Central America." Presented at Workshop on Fermentation Alcohol for Use as Fuel and Chemical Feedstock in Developing Countries; 26-30 March 1979; Vienna, Austria. Sponsored by the United Nations Industrial Development Organization.

Karan, Ram. 1979. "Cane Molasses Fermentation Alcohol Industry in Fiji." Presented at Workshop on Fermentation Alcohol for Use as Fuel and Chemical Feedstock in Developing Countries; 26-30 March 1979; Vienna, Austria. Sponsored by the United Nations Industrial Development Organization.

Knapps, Carol. 1981. Adviser to the Executive Director, Inter-American Development Bank, Washington, D.C. Personal communication.

Kohli, Harinder. 1980. Central Industrialized Projects Unit, The World Bank, Washington, D.C. Personal communication.

Lima, Sergio E. Moreira. 1980. Energy Office, Embassy of Brazil, Washington, D.C. Personal communication.

Liquid Fuels Trust Board. 1979. Report of the Liquid Fuels Trust Board of New Zealand to Parliament. Wellington, New Zealand: LFTB.

Madi, A. G. 1979. "Fermentation Alcohol Industry in Egypt in the Last Three Decades." Presented at Workshop on Fermentation Alcohol for Use as Fuel and Chemical Feedstock in Developing Countries; 26-30 March 1979; Vienna, Austria. Sponsored by the United Nations Industrial Development Organization.

Mungai, J. J. 1979. "Molasses Production and Utilization Potential in Tanzania." Presented at Workshop on Fermentation Alcohol for Use as Fuel and Chemical Feedstock in Developing Countries; 26-30 March

234

1979; Vienna, Austria. Sponsored by the United Nations Industrial Development Organization.

Nagata, Eiichi. 1980. Director, Agriculture Department of the Japan Trade Center, Chicago, IL. Personal communication.

New Zealand Energy Research and Development Committee. 1980. "The Potential of Energy Fairness for Transport Fuels in New Zealand." Report No. 46. Wellington, New Zealand: ERDC.

Sangster, I. 1979. "The Potential of Sugar Cane Derived Alcohol as a Fuel in Jamaica." Presented at Workshop on Fermentation Alcohol for Use as Fuel and Chemical Feedstock in Developing Countries; 26-30 March 1979; Vienna, Austria. Sponsored by the United Nations Industrial Development Organization.

Sharma, K.D. 1979. "Present Status of Alcohol and Alcohol Based Chemicals Industry in India." Presented at Workshop on Fermentation Alcohol for Use as Fuel and Chemical Feedstock in Developing Countries; 26-30 March 1979; Vienna, Austria. Sponsored by the United Nations Industrial Development Organization.

Swedish Commission for Oil Substitution. 1980. "Introduction of Alternative Motor Fuels." Stockholm, Sweden: Swedish Commission for Oil Substitution.

Tatum, John. 1980. Energy Consultant, Smyrna, GA. Personal communication.

Yokoi, A. 1979. "Consumption Figures of Fermentation Alcohol in Japan." Presented at Workshop on Fermentation Alcohol for Use as Fuel and Chemical Feedstock in Developing Countries; 26-30 March 1979; Vienna, Austria. Sponsored by the United Nations Industrial Development Organization.

Appendix:
Technical Reference Data

Area

1 acre = 43,560 square feet; 4,840 square yards; 0.40
    hectares
1 hectare = 10,000 square meters; 2.47 acres

Heat, Energy

1 Btu = 252 calories, which is the heat required to
    raise 1 pound of water 1 degree Fahrenheit
Btu/barrel from crude oil = 5,800,000
Btu/barrel from motor gasoline = 5,253,000
Btu/barrel from natural gasoline = 4,620,000
1 calorie = .00397 Btu, which is the heat required to
    raise 1 gram of water 1 degree Celsius
1 quad = 1 quadrillion (10^{15}) Btu;
To convert from °F to °C, subtract 32 and then divide by
    1.8
To convert from °C to °F, multiply by 1.8 and then add
    32

Mass, Weight, Volume

1 barrel of crude oil = 42 gallons; 0.136 metric tons;
    0.150 short tons
1 bushel = 8 gallons
1 cubic foot = 7.48 liquid gallons; 62.36 H_2O (at 60°F)
1 gallon of 200-proof ethanol = 6.6 pounds
1 liter = 1.057 U.S. liquid quarts; 0.908 dry quarts;
    61.02 cubic inches
1 metric ton = 1,000 kilo grams; 1.1 U.S. tons; 7.33
    barrels
1 pound = 453.6 grams
1 short ton = 6.05 barrels; 2,000 pounds
1 U.S. liquid gallon water = 8.33 pounds (at 60°F);
    0.134 cubic foot; 231 cubic inches; 128 fluid
    ounces; 4 quarts; 8 pints; 3.785 liters

CHEMICAL FORMULAS

Ammonia	–	NH_3
Butanol	–	C_4H_9OH
Calcium oxide	–	C_aO
Carbon dioxide	–	CO_2
Carbon monoxide	–	CO
Ethanol	–	C_2H_5OH, also EtOH
Ethylene glycol	–	H_4C_22OH
Fructose	–	$C_6H_{12}O_6$
Gas	–	CH_2
Glucose	–	$C_6H_{12}O_6$
Glycerol	–	C_3H_53OH
Hydrated lime	–	$C_a(OH)_2$
Hydrochloric acid	–	HC_1
Hydrogen	–	H_2
Isopropyl alcohol	–	C_3H_7OH
Lactic acid	–	$C_3H_6O_3$
Lactose	–	$C_{12}H_{22}O_{11}$
Lime	–	C_aO
Methane	–	CH_4
Methanol	–	CH_3OH
Methyl tert-butyl ether	–	$(CH_3)_3COCH_3$
Nitric oxide	–	NO
Nitrogen	–	N_2
Nitrogen oxides	–	NO_x
Nitrous oxide	–	N_2O
Octane	–	C_8H_{18}
Oxygen	–	O_2
Sucrose	–	$C_{12}H_{22}O_{11}$
Sulphuric acid	–	H_2SO_4
Tert-butyl alcohol	–	$(CH_3)_3COH$

ABBREVIATIONS

After top dead center	AFDC
American Chemical Society	ACS
American National Safety Institute	ANSI
American Society of Mechanical Engineers	ASME
American Society of Testing and Materials	ASTM
Atmosphere	atm
Australian dollars	$A
Before top dead center	BTCD
Biological oxygen demand	BOD
British thermal unit	Btu
Bureau of Alcohol, Tobacco and Firearms	BATF or ATF
Bushel	bu
Calorie	cal
Commodity Credit Corporation	CCC
Cubic foot	ft^3
Cubic meter	m^3
Degree Celsius	^oC
Degree Fahrenheit	^oF
Degree Kelvin	K
Department of Agriculture	USDA
Department of Labor	USDOL
Department of Transportation	USDOT
Distillers dried grains, distillers dark grains	DDG
Distillers dried grains with solubles	DDGS
Distillers dried solubles	DDS
Economic Development Act	EDA
Ethanol	EtOH
Farmer's Home Administration	FmHA
Federal testing procedure	FTP
Fiscal year	FY
Gross national product	GNP
Hectare	ha
Hundredweight	cwt
Internal combustion	IC
Joules	J
Kenya shilling	K sh
Kilo Pascale	kPa
Kilogram	kg
Kilojoule	kJ
Kiloliter	kl
Kilowatt	kW
Kilowatt hour	kWh
Liter	L
Maximum velocity of a reaction	V_{max}

```
Megawatt . . . . . . . . . . . . . . . . . . . . . MW
Meter . . . . . . . . . . . . . . . . . . . . . . . m
Methyl tert-butyl ether . . . . . . . . . . . . . MTBE
Millimolar . . . . . . . . . . . . . . . . . . . . mM
Million Btu . . . . . . . . . . . . . . . . . . . MBtu
Mole percent . . . . . . . . . . . . . . . . . . mol %
Molecular weight . . . . . . . . . . . . . . . . mol wt
Motor Octane Number . . . . . . . . . . . . . . . MON
National Alcohol Fuel Commission . . . . . . . . NAFC
National Fire Protection Association . . . . . . NFPA
Occupational Safety and Health
  Administration . . . . . . . . . . . . . . . . OSHA
Parts per billion . . . . . . . . . . . . . . . . ppb
Parts per million . . . . . . . . . . . . . . . . ppm
Quad . . . . . . . . . . . . . . . . . . . . . . . Q
Research Octane Number . . . . . . . . . . . . . RON
Simultaneous saccharification and fermentation . . SSF
Solar Energy Research Institute . . . . . . . . . SERI
Spark ignition . . . . . . . . . . . . . . . . . . SI
Tert-butyl alcohol . . . . . . . . . . . . . . . TBA
Top dead center . . . . . . . . . . . . . . . . . TDC
Ultrafiltration . . . . . . . . . . . . . . . . . UF
United States Department of Energy. . . . USDOE also DOE
Universal quasi-chemical . . . . . . . . . . . UNIQUAC
Volume percent . . . . . . . . . . . . . . . . . vol %
Weight percent . . . . . . . . . . . . . . . . . wt %
```

Glossary

ABSORPTION: taking up of one substance by another to its inner structure.

ACETALDEHYDE: a low-boiling industrial chemical used in the synthesis of other compounds.

ACID HYDROLYSIS: decomposition or alteration of a chemical substance by means of an acid.

ACIDITY: the measure of how many hydrogen ions a solution contains.

ADSORPTION: taking up of one substance by another to its surface.

AFLATOXIN: a substance produced by certain strains of the fungus Aspergillus flavus, which contaminates corn and is a carcinogenic.

ALCOHOL: the family name of a group of organic chemicals comprising carbon, hydrogen, and oxygen, which are differentiated by their number of carbon molecules.

ALDEHYDES: a group of highly reactive organic chemical compounds obtained by oxidation of the primary alcohols, and characterized by their carbon, hydrogen, oxygen (CHO) group. They are used in the manufacture of resins, dyes, and organic acids.

ALPHA-AMYLASE: an enzyme which converts starch into sugars.

AMINO ACIDS: the naturally occurring, nitrogen-containing building blocks of protein.

ANAEROBIC DIGESTION: a type of bacterial degradation of organic matter that occurs only in the absence of air (oxygen).

239

ANHYDROUS: a compound that does not contain water.

APPROPRIATE TECHNOLOGY: the technology necessary to meet a group's, community's, or region's commonly identified development needs; the determining factor is the identified priorities and conditions, not the size or complexity of the technology.

ATMOSPHERIC PRESSURE: the pressure of the air (and the surrounding atmosphere), which changes from day to day.

AUTOIGNITION: self-ignition in an internal combustion engine either from the heat of compression or a spark.

AZEOTROPE: the chemical term for two liquids that, at a certain concentration, vaporize at the same temperature; alcohol and water cannot be separated further than about 95% alcohol-5% water because at this concentration, alcohol and water form an azeotrope and vaporize together.

BAGASSE: the crushed remains of sugarcane after the juice has been extracted; may be used as a fuel for process heat, as a source of carbohydrate for processes such as paper production, or as an addition to animal feeds.

BATCH FERMENTATION: fermentation conducted from start to finish in a single vessel.

BEER: the product of fermentation by microorganisms; the fermented mash, which contains about 11% - 12% alcohol; usually refers to the alcohol solution remaining after yeast fermentation of sugars.

BEER STILL: the stripping section of a distillation column for concentrating ethanol.

BENZENE: C_6H_6; a liquid, aromatic, colorless, volatile, flammable hydrocarbon.

BIOMASS: plant and animal matter, wastes, and residues that have the potential of being converted into useable energy resources, for example, crops, wood, and sewage.

BRITISH THERMAL UNIT (Btu): the amount of heat required to raise the temperature of one pound of water one degree Fahrenheit under stated conditions of pressure and temperatures (equal to 252 calories, 778 foot-pounds, 1,055 joules, and 0.293 watt-hours); it is the standard unit for measuring quantities of heat energy.

CALORIE: the amount of heat required to raise one gram of water one degree celsius.

CARBOHYDRATE: a group of neutral compounds comprising carbon, hydrogen, and oxygen including sugars, starches, and celluloses.

CASSAVA: a plant of the genus Manihot with a fleshy root, grown in the tropics for food; also called manioc and tapicoa plant.

CELLULOSE: a natural carbohydrate polysaccharide with long, essentially linear molecular chains.

CETANE NUMBER: the measure of the ignition value of diesel fuel.

COLUMN: a vertical, cylindrical vessel used to increase the degree of separation of liquid mixtures by distillation or extraction.

COMPOUND: a chemical term denoting a combination of two or more distinct elements.

CONCENTRATION: the ratio of mass or volume of solute present in a solution to the amount of solvent.

CONDENSER: a heat-transfer device that reduces a thermodynamic fluid from its vapor phase to its liquid phase.

CONTINUOUS FERMENTATION: a steady-state fermentation system that operates without interruption; each stage occurs in a separate section of the fermentor, and flow rates are set to correspond with required residence times.

DENATURATION: the process of adding a substance to ethanol to make it unfit for human consumption; in the United States, the denaturing agent may be gasoline or other substances specified by the Bureau of Alcohol, Tobacco and Firearms.

DESICCANT: a substance having an affinity for water; used for drying purposes.

DESTRUCTIVE DISTILLATION: a process in which high cellulose-content organic wastes, such as wood wastes, are heated in the absence of oxygen to decompose them, and then are distilled to produce methanol.

DEXTRINIZATION: see LIQUEFACTION.

DEXTRIN: a polymer that is intermediate in complexity between starch and maltose, and is formed by hydrolysis of starches.

DISACCHARIDE: the class of compound sugars that yield two monosaccharide units upon hydrolysis; examples are sucrose, mannose, and lactose.

DISTILLATE: that portion of a liquid that is removed as a vapor and condensed during the distillation process.

DISTILLATION: the process of separating the components of a mixture by using their differences in boiling point. A vapor is formed by heating the liquid in a vessel; the vapors are then successively collected and condensed into liquids.

DISTILLERS DARK GRAINS: DDG; the mixture which results when distillers dried grains (also called DDG) are combined with the solubles obtained from fermentation stillage. This product is similar to DISTILLERS DRIED GRAINS WITH SOLUBLES.

DISTILLERS DRIED GRAINS: DDG; the coarse-grain fraction of fermentation stillage, which is separated from the liquid and dried.

DISTILLERS DRIED GRAINS WITH SOLUBLES: DDGS; the mixture remaining after condensing and drying at least three-fourths of the solids in fermentation stillage. This product is similar to DISTILLERS DARK GRAINS.

DISTILLERS DRIED SOLUBLES: DDS; the condensing and drying of the thin stillage, which is the water-soluble fraction of fermented mash plus the mash water.

DISTILLERS GRAIN: the nonfermentable portion of a grain mash comprising protein, unconverted carbohydrates and sugars, and inert material.

ENSILAGE: See SILAGE.

ENZYMES: the group of catalytic proteins that are produced by living microorganisms; enzymes mediate and promote the chemical processes of life without themselves being altered or destroyed.

ETHANOL: C_2H_5OH; the alcohol product of fermentation that is used in alcohol beverages and for industrial purposes; the alcohol blended with gasoline to make gasohol and other similar belnds; also known as ethyl alcohol or grain alcohol.

ETHYL ALCOHOL: also known as ethanol or grain alcohol; see ETHANOL.

ETHYLENE: a colorless gas derived from thermal cracking of hydrocarbons or dehydration of ethanol.

FAHRENHEIT: a temperature scale in which the boiling point of water is 212° and its freezing point is 32°; to convert °F to °C, subtract 32, then divide by 1.8.

FEEDSTOCK: the base raw material for industrial processes, in the case of ethanol fermentation it consists of monomeric or other fermentable sugars derived from specific crops.

FERMENTABLE SUGAR: sugar (usually a monomeric sugar such as glucose) derived from starch and cellulose that can be converted to ethanol; also known as reducing sugar or monosaccharide. See also MONOSACCHARIDE.

FERMENTATION: a microorganically mediated enzymatic transformation of organic substances, especially carbohydrates, generally accompanied by the production of a gas.

FERMENTATION ETHANOL: ethanol produced from enzymatic transformation of organic substances.

FLASH POINT: the temperature at which a combustible liquid will ignite when a flame is introduced.

FLOCCULATION: the aggregation of fine suspended particles which form floating clusters or clumps.

FOSSIL FUEL: a fuel created by the compression of organic matter from prior geologic eras in the earth's crust, e.g., coal, crude oil, natural gas.

FRACTIONAL DISTILLATION: a process of separating alcohol and water (or other mixtures).

FRUCTOSE: $C_6H_{12}O_6$; a fermentable monosaccharide or simple sugar. Fructose and glucose are optical isomers, that is, their chemical structures are the same, but their geometric configurations are mirror images of one another.

FUEL CELLS: a means for converting a fuel and an oxidant into electricity unlike conventional batteries in that the fuel is fed to the cell as needed.

FUSEL OIL: a clear, colorless, toxic liquid mixture of alcohols obtained as a by-product of grain fermentation;

generally amyl, iso-amyl, n-propyl, and iso-butyl
alcohols, and acetic and lactic acids.

GASOHOL: registered trade name by the Nebraska
Agricultural Industrial Utilization Committee for a
blend of 90% unleaded gasoline with 10% agriculturally
derived fermentation ethanol.

GASOLINE: a volatile, flammable liquid containing a
mixture of hydrocarbons obtained from petroleum that has
a boiling range of approximately 29° - 216°C and is used
as fuel for spark-ignition internal combustion engines.

GELATINIZATION: a process in which starch granules are
heated, causing them to rupture and form a gel of
soluble starch and dextrins.

GLUCOSE: $C_6H_{12}O_6$; a monosaccharide and the most common
sugar; an optical isomer of fructose.

GLYCOLYSIS: enzymatic decomposition of carbohydrates,
which releases energy; occurs during yeast fermentation.

GRAIN ALCOHOL: see ETHANOL.

HEATING VALUE: the amount of heat obtainable from a
fuel and expressed in Btu per unit measurement of the
fuel.

HEMICELLULOSE: beta and gamma cellulose; cellulose with
a degree of polymerization of $\leq$ 150.

HEXOSE: any of various monosaccharides with six carbon
atoms in the molecule.

HYDRATED: chemically combined with water.

HYDROCARBON: a chemical compound containing hydrogen,
oxygen, and carbon.

HYDROLYSIS: the alteration of a compound into other
compounds by the addition of a water molecule.

INDOLENE: a petroleum fuel used in comparative tests of
automotive fuels.

INOCULUM: a small amount of bacteria produced from a
pure culture that is used to start a new culture.

INULIN: a polymeric carbohydrate made up of fructose
monomers found in many plants, including the Jerusalem
artichoke.

JOULE: the absolute unit of work or energy equal to 10^7 ergs, or approximately 0.7375 foot-pounds, or 0.2390 gram-calories.

KILO PASCALE: a unit of pressure measurement.

LACTOSE: a white disaccharide derived from milk products such as whey; also called milk sugar.

LEAN FUEL MIXTURE: an excess of air in the air-fuel ratio.

LIGNIN: a substance that with cellulose forms the woody cell walls of plants. It is a polymeric material characterized by a higher carbon content than cellulose, and by propyl-benzene units, methoxyl groups, and hydroxyl groups.

LIGNIFIED CELLULOSE: a cellulose polymer wrapped in a polymeric sheath with linkages called lignin, which makes it resistant to hydrolysis.

LIQUEFACTION: the process of changing a substance into the liquid state; in fermentation, specifically converting water-insoluble carbohydrate to water-soluble carbohydrate.

MALT: barley softened in water and allowed to germinate; used especially in brewing and distilling.

MASH: a mixture of any grain feedstock and other ingredients with water in preparation for fermentation.

METHANE: CH_4; a colorless gas of the paraffin hydrocarbon series; also called marsh gas or methyl hydride.

METHANOL: CH_3OH; the simplest alcohol; formed by the destructive distillation of wood, from biomass, or from coal gasification. It is used as an antifreeze, a solvent, a fuel, and in production of other chemicals; also known as methyl alcohol and wood alcohol.

METHYL ALCOHOL: see METHANOL.

MICHAELIS CONSTANT: the concentration of substrate at which the speed of the reaction is one-half of maximum velocity.

MILLIMOLAR: the number of millimoles of substrate per liter in a reaction.

MISCIBLE: capable of being mixed in any ratio without separation.

MOLECULAR SIEVE: a column which separates molecules by selective adsorption of molecules on the basis of size.

MOLECULE: the chemical term for the smallest particle of matter that is the same chemically as the whole mass.

MONOMER: a simple compound which is capable of combining with other like or unlike molecules to form a polymer.

MONOSACCHARIDE: any simple sugars with the formula $C_6H_{12}O_6$; fermentable sugars, such as glucose, derived from starch, cellulose, or other sugars.

MOTOR OCTANE NUMBER: a numerical rating measuring the tendency of an engine to knock when a motor fuel is used in road operating conditions.

NEAT FUELS: pure or straight fuels.

NEUTRAL SOLVENTS: non-polar substances capable of dissolving another substance to form a solution.

OCTANE NUMBER: a numerical rating which indicates the tendency of an engine to knock when a motor fuel is used in a standard spark-ignition internal combustion engine. See also MOTOR OCTANE NUMBER, RESEARCH OCTANE NUMBER, and PUMP OCTANE NUMBER.

OSMOSIS: the tendency of a liquid to pass through a semipermeable membrane into a solution where its concentration is lower, thus equalizing conditions on either side of the membrane.

OSMOTIC PRESSURE: the pressure necessary to prevent passage of a solvent across a membrane (which could be a cell wall), thereby separating solutions of different concentrations.

PETRANOL: a blend of 10% ethanol and 90% supergrade petrol, which is being test marketed by the Shell Co. in Australia.

pH: a term used to describe the free hydrogen-ion concentration of a solution; a solution of pH 0 to <7 is acid, 7 is neutral, >7 to 14 is alkaline.

POLYMER: a substance made of long chained molecules or cross-linked simple molecules.

POLYSACCHARIDE: a carbohydrate decomposable by hydrolysis into two or more molecules of monosaccharides or their derivatives.

PROOF: a measure of alcohol content; one percent equals 2 proof.

PROOF GALLON: a U.S. gallon of liquid at 15.5°C that is 50% ethanol by volume.

PROTEIN: any of a class of nitrogenous high molecular-weight polymer compounds that yield amino acids required in animal metabolism to carry out life processes.

PUMP OCTANE NUMBER: the Research Octane Number plus the Motor Octane Number divided by two. See also MOTOR OCTANE NUMBER and RESEARCH OCTANE NUMBER

PYROLYSIS: the breaking apart of complex molecules into simpler units by heating in the absence of oxygen.

QUAD: one quadrillion (10^{15}) Btu.

RECTIFICATION: in distillation, the selective increase of a concentration of the lower volatile component in a mixture by successive evaporation and condensation.

RECTIFYING COLUMN: the portion of a distillation column above the feed tray in which rising vapor interacts with a countercurrent falling stream of condensed vapor.

REFLUX: the portion of the product stream which is returned to the process to assist in increasing conversion or recovery.

RENEWABLE RESOURCE: a resource, such as for energy use, which is continually replaced through natural means; e.g., wind, solar, and biomass.

RICH FUEL MIXTURE: an excess of fuel in the air-fuel ratio.

RESEARCH OCTANE NUMBER: a numerical rating measuring the tendency of an engine to knock when a motor fuel is used in standard test conditions.

RUMINANT: a cud-chewing animal.

SACCHARIDE: a simple sugar, combination of sugars, or polymerized sugar, e.g., monosaccharide, disaccharide, or polysaccharide.

SACCHARIFICATION: the process of changing or
hydrolyzing a complex carbohydrate into a simpler
soluble fermentable or monomeric sugar, such as glucose.

SACCHAROMYCES: a class of single-cell yeasts which
selectively consume simple sugars.

SILAGE: fodder, such as field corn or hay, which has
undergone an acid fermentation to retard spoilage, used
as a livestock feed. Also called ensilage.

SLURRY: a mixture of water and insoluble matter.

SOLVENT: a substance that can dissolve another
substance and form a solution.

STARCH: a carbohydrate polymer consisting of glucose
monomers linked together in a particular pattern (a
glycosidic bond organized in repeating units), which is
found in most plants, and is the principal energy-
storage form of photosynthesis.

STILL: an apparatus for distilling liquids,
particularly alcohols, consisting of a vessel in which
the liquid is vaporized by heat and a cooling device in
which the vapor is condensed.

STILLAGE: the nonfermentable residue, both solids and
liquids, from the fermentation of mash to produce
alcohol.

STOICHIOMETRIC: the exact calculation of the chemical
elements or compounds necessary for a particular
reaction to occur.

STOVER: the dried stalks and leaves of a crop remaining
after the grain has been harvested.

STRATIFIED CHARGE: a variation to a spark-ignition
internal combustion engine with a divided chamber, in
which rich air-fuel mixture is injected, ignited so that
combustion begins, and then fed into the second chamber,
where it is sparked again and combustion occurs.

STREPTOMYCES: a genus of the family Streptomycetaceae,
which are higher bacteria.

SUCROSE: $C_{12}H_{22}O_{11}$; a crystalline disaccharide
carbohydrate found in some plants, mainly sugarcane,
sugar beets, and maple trees.

SUBSTRATE: a substance acted upon in a reaction; a
source of reactive material.

SYNFUEL: a nontechnical term popularly used to refer to a variety of nonpetroleum fuels, e.g., shale oil fuels, coal gasification fuels.

SYNGAS: a gaseous mixture used for synthesizing various organic and inorganic compounds, also called synthesis gas.

THERMOPHILIC: capable of growing and surviving at high temperatures.

TURBINE: a rotary engine actuated by a flow of fluid or vapor under pressure, which moves a series of vanes located on a central core, enclosed by a casing.

VAPOR LOCK: an interruption of fuel flow in an internal combustion engine caused by vapor or gas bubbles in the fuel-feed system.

VAPORIZE: to change from a liquid or a solid to a vapor, as in heating water to steam.

WHEY: the watery part of milk separated from the curd in the process of making cheese.

WORT: the liquid remaining from a brewing mash preparation following the filtration of fermentable beer.

YEAST: a single-cell microorganism (fungi) that contains enzymes capable of changing sugar to alcohol by fermentation.

Selected Annotated
Bibliography

BUTANOL

Beesch, Samuel C. "Acetone-Butanol Fermentation of Sugars." Industrial and Engineering Chemistry. Vol. 44, no. 7, July 1952.

This article contains information on the microorganisms used for butanol-acetone fermentation and detailed information on the fermentation process.

Compere, A.L., Griffith, W.L., Googin, J.M. Fuel Alcohol Extraction Technology Commercialization Conference. Oak Ridge, TN: Oak Ridge National Laboratory; 1980. Document no. CONF-801212. [Available from the National Technical Information Service, 5285 Port Royal Road, Springfield, VA 22161.]

A thorough description of Oak Ridge National Laboratory's fuel alcohol extraction (fualex) process is given in this report.

Prescott, Samuel C., Dunn, Cecil G. Industrial Microbiology. New York, NY: McGraw-Hill Book Company, Inc.; 1949.

Chapters on butanol-acetone fermentation and butanol-isopropyl alcohol fermentation are included, as well as general discussions on yeasts and saccharification.

Rose, Anthony H. Industrial Microbiology. Washington, D.C.: Butterworth and Co., Ltd.; 1961.

Extensive information is given on industrial microbiology, as well as a chapter on the butanol-acetone fermentation process.

Strobel, M.K., Bader, J.B. Economic Evaluation of
Neutral-Solvents Fermentation Product Separation. Oak
Ridge, TN: Oak Ridge Station, School of Chemical
Engineering Practice, Massachusetts Institute of
Technology and Oak Ridge National Laboratory; 1981.
Document no. ORNL/MIT-330. [Available from the National
Technical Information Service, 5285 Port Royal Road,
Springfield, VA 22161.]

This paper compares the cost of butanol-acetone
fermentation using Clostridia as the microorganism with
ethanol fermentation. Market costs as well as
production costs are included in the calculations.

Underkofler, L.D., Hickey, R.J., eds. Industrial
Fermentations. Vol. 1. New York, NY: Chemical
Publishing Co.; 1954.

This book serves as a basic chemical text on
fermentation processes. It contains a chapter on
butanol-acetone fermentation.

ETHANOL

Distillers Feed Research Council. Distillers Feeds.
Cincinnati, OH: Distillers Feed Research Council; n.d.
[Available from Distillers Feed Research Council, 1435
Enquirer Bldg., Cincinnati, OH 45202.]

This pamphlet provides a well-documented and referenced
summary of the use of stillage as animal feed.

Distillers Feed Research Council. Feed Formulation.
Cincinnati, OH: Distillers Feed Research Council; n.d.
[Available from Distillers Feed Research Council, 1435
Enquirer Bldg., Cincinnati, OH 45202.]

The purpose of this pamphlet is to provide information
on distillers dried grains, distillers dried grains with
solubles, distillers dried solubles, and condensed
distillers solubles in order to produce better formula
feeds for animals.

Lipinsky, E.S., et al. Sugar Crops as a Source of
Fuels. Vol. I-Agricultural Research, Vol. II-Processing
and Conversion Research. Washington, D.C.: U.S.
Department of Energy; 1978. Report no. TID-29400/1 and
TID-29400/2. [Available from the National Technical
Information Service, 5285 Port Royal Road, Springfield,
VA 22161.]

The two volumes of this report present the results of a feasibility study on using sugarcane and sweet sorghum as a source of fuels. Field experiments were conducted in Florida, Louisiana, Mississippi, Ohio, and Texas.

Moriarity, Andrew J. "Toxicological Aspects of Alcohol Fuel Utilization." Toronto, Canada: Biomedical Resources International; November 1977.

Dr. Moriarity's paper presents a full discussion on the toxicology of ethanol and methanol and a comparison with the toxicity of gasoline.

Office of Technology Assessment. Gasohol-A Technical Memorandum. Washington, D.C.: Office of Technology Assessment, U.S. Congress; 1979. [Available from Superintendent of Documents, U.S. Government Printing Office, Stock no. 052-003-00706.]

This document discusses ethanol production and gasohol use as well as economics, environmental effects, social impacts, and federal programs and policies.

Paturau, J.M. By-products of the Cane Sugar Industry. Amsterdam, The Netherlands: Elsevier Publishing Company; 1969.

As well as a complete discussion on fermentation ethanol production from sugarcane, this book provides a comprehensive examination of those coproducts of ethanol fermentation when molasses is used as the feedstock.

Paul, J.J., ed. Ethyl Alcohol Production and Use as a Motor Fuel. Park Ridge, NJ: Noyes Data Corp.; 1979.

This book presents information from other sources on economic assessments of ethanol production from biomass, feedstock availability, ethanol production technology, and use of ethanol and ethanol-gasoline blends.

Solar Energy Research Institute. Fuel From Farms: A Guide to Small-Scale Ethanol Production. Golden, CO: Solar Energy Research Institute; 1980. Report no. SERI/SP-451-519. [Available from the National Technical Information Service, 5285 Port Royal Road, Springfield, VA 22161.]

Invaluable information on the potential of small-scale, on-farm ethanol production is given in this guide. Useful decison and planning worksheets and a business-plan case study are included.

Solar Energy Research Institute. A Guide to
Commercial-Scale Ethanol Production and Financing.
Golden, CO: Solar Energy Research Institute; 1980.
Report no. SERI/SP-751-877. [Available from the
National Technical Information Service, 5285 Port Royal
Road, Springfield, VA 22161.]

This guide is the commercial-scale equivalent of Fuel
From Farms. Information for determining feasibility is
included.

Underkofler, L.D., Hickey, R.J., eds. Industrial
Fermentations. Vol. 1. New York, NY: Chemical
Publishing Co.; 1954.

This book serves as a basic chemical text on
fermentation processes. It includes chapters on
fermentation of grain, molasses, sulphite waste liquor,
and wood waste, as wella as yeast production.

U.S. Department of Agriculture. Small-Scale Fuel
Alcohol Production. Washington, D.C.: U.S. Department
of Agriculture; 1980.

This book provides information on small-scale, on-farm
ethanol production. The emphasis is on utilization of
alcohol fuel in vehicles and use of the stillage
by-product. Additional information is given on
feedstocks, production processes, and costs.

GENERAL

Douglas, Larry. The Chemistry and Energetics of
Biomass Conversion, an Overview. Golden, CO: Solar
Energy Institute; 1981.

This paper provides an overview of biomass energy
systems and a discussion of the research on the
chemistry and structures of lignocellulosic biomass,
biomass thermal conversion, and biochemical conversion.

Pleeth, S.J.W. Alcohol, a Fuel in Internal
Combustion Engines. London, England: Chapman and Hall;
1949.

Written in 1949, this book represents the classic text
on alcohol fuels. Today, it still contains valuable
information on production and use of alcohol fuels.

Solar Energy Research Institute. Proceedings from the Third Annual Biomass Energy Systems Conference Proceedings. Golden, CO: Solar Energy Research Institute; 5 - 7 June 1979. Report no. SERI/IP-33-285. [Available from the National Technical Information Service, 5285 Port Royal Road, Springfield, VA 22161.]

This document contains over 70 individual papers on a wide range of subjects concerned with biomass energy systems with emphasis on production of ethanol and methanol.

United Nations Industrial Development Organization. Workshop on Fermentation Alcohol for Use as a Fuel and Chemical Feedstock in Developing Countries. Vienna, Austria; 6-30 March 1979.

The papers presented at this workshop focus on historical and current production and use of alcohol fuels in the developing countries. The emphasis is on use of alcohols as chemical feedstocks.

Volkswagenwerk AG, ed. Proceedings of the International Symposium on Alcohol Fuel Technology - Methanol and Ethanol. Wolfsburg, Federal Republic of Germany; 21-23 November 1977. Washington, D.C.: U.S. Department of Energy; 1978. Report no. CONF 771175.

This collection of papers presents invaluable information on methanol and ethanol production and use. Over 40 papers are included concerning alcohol-fuel technology. Subjects discussed include economic and political implications, use as a vehicle fuel, production from different feedstocks, environmental issues, and optimization of alcohol-fuel use.

METHANOL

Hagen, David L. "Methanol as a Fuel: A Review with Bibliography." Warrendale, PA: Society of Automotive Engineers; 1977. Report no. SAE/PT-80/19.

This paper provides historical information on methanol production and use and is an excellent survey of recent studies and research. A comprehensive survey of methanol use is also given.

Moriarity, Andrew J. "Toxicological Aspects of Alcohol Fuel Utilization." Toronto, Canada: Biomedical Resources International; November 1977.

Dr. Moriarity's paper presents a full discussion on the toxicology of ethanol and methanol and a comparison with the toxicity of gasoline.

Paul, J.K. Methanol Technology and Application in Motor Fuels. Park Ridge, NJ: Noyes Development Corp.; 1978.

The information presented in this book is derived from a variety of sources and covers the production of methanol from coal, solid waste, and natural gas. Reports on the use of straight methanol and methanol-gasoline blends as motor vehicle fuels are discussed. The Mobil process for producing gasoline from methanol is included also.

Reed, T.B. "Net Efficiencies of Methanol Production from Gas, Coal Waste or Wood." Symposium on Net Energetics of Integrated Synfuel Systems. 171st National Meeting of the American Chemical Society, Division of Fuel Chemistry. New York, NY: American Chemical Society; April 1976. Vol. 21, no. 2, paper no. 16.

Dr. Reed's paper discusses and compares the net production energy efficiencies for methanol produced from gas, coal, waste, and wood.

POLICY ISSUES

Berry, Wendell. The Unsettling of America: Culture and Agriculture. New York, NY: Avon Books; 1978.

This book provides a framework for understanding the importance of the small-scale, family-owned, integrated farm in America today. It also critiques modern agricultural policy.

Brown, Lester R. "Food or Fuel; New Competition for the World's Cropland." Washington, D.C.: Worldwatch Institute; 1980.

Mr. Brown presents an argument that there is not enough usable land for food and fuel production.

Flanagan, Dennis, ed. Scientific American. Vol. 243, no. 3. New York, NY: Scientific American, Inc.; September 1980.

This particular issue of Scientific American is devoted to economic development. It presents an excellent overview of the global situation on economic development, population, food, energy, and water. Case studies of China, India, Tanzania, and Mexico are presented, as well as a model of the world economy in the year 2000.

Lappe, Frances Moore, Collins, Joseph. Food First: Beyond the Myth of Scarcity. New York, NY: Ballantine Books; 1979.

The focus of this book is a comprehensive and thoroughly referenced account of food production and distribution in the world today. The basic conclusion is that availability of food is more closely related to politics than to lack of arable land or other necessary resources.

Lockeretz, W., et al. "Maize Yields and Soil Nutrient Levels With and Without Pesticides and Standard Commercial Fertilizers." Agronomy Journal. Vol. 72, January-February 1980. Pp. 65-72.

This article reports on a study comparing two groups of 26 commercial mixed grain and livestock farms in the western corn belt. One group used conventional fertilizers and pesticides, the other used none. Because of rising costs and uncertain supplies for agricultural chemicals, it is important to establish data on production output that presents alternatives. This study found that the difference was not statistically significant and that conventional yields were higher under favorable growing conditions and lower in unfavorable conditions.

U.S. National Alcohol Fuels Commission. Alcohol Fuels Tax Incentives-A Summary: Alcohol Fuels Provisions of the Crude Oil Windfall Profit Tax Act. Washington, D.C.: U.S. National Alcohol Fuels Commission; 1980. This booklet reviews those provisions in the Crude Oil Windfall Profit Tax Act relevant to alcohol fuels including: four-cent-per-gallon excise tax exemption, income tax credits, energy investment tax credit for biomass, alcohol fuel plant operating permits, tax-exempt bonds for alcohol fuel from solid waste, state financing of renewable energy property, imported alcohol study, and the annual reports on alcohol fuels.

As well as a complete discussion on production, this book provides a comprehensive examination of those coproducts of ethanol fermentation when molasses is used as the feedstock.

U.S. House of Representatives, Ninety-fifth Congress,
Second Session. Alcohol Fuels: Hearings Before the
Subcommittee on Advanced Energy Technologies and Energy
Conservation, Research, Development and Demonstration of
the Committee on Science and Technology. Washington,
D.C.: U.S. House of Representatives; 11-12 July 1978.
(Available from Superintendent of Documents. U.S.
Government Printing Office, Washington, D.C. 20402.)

Testimony at these hearings present a variety of views
on the potential of alcohol fuels by representatives of
private industry and government. Detailed reports with
bibliographies are included as well as submitted
testimony statements.

Index

264